Recent Vitamin R

Editor

Michael H. Briggs, D.Sc., Ph.D.

Professor of Human Biology
Dean of Sciences
Deakin University
Geelong, Victoria, Australia

CRC Press
Taylor & Francis Group
Boca Raton London New York

CRC Press is an imprint of the
Taylor & Francis Group, an **informa** business

First published 1984 by CRC Press
Taylor & Francis Group
6000 Broken Sound Parkway NW, Suite 300
Boca Raton, FL 33487-2742

Reissued 2018 by CRC Press

© 1984 by Taylor & Francis
CRC Press is an imprint of Taylor & Francis Group, an Informa business

No claim to original U.S. Government works

This book contains information obtained from authentic and highly regarded sources. Reasonable efforts have been made to publish reliable data and information, but the author and publisher cannot assume responsibility for the validity of all materials or the consequences of their use. The authors and publishers have attempted to trace the copyright holders of all material reproduced in this publication and apologize to copyright holders if permission to publish in this form has not been obtained. If any copyright material has not been acknowledged please write and let us know so we may rectify in any future reprint.

Except as permitted under U.S. Copyright Law, no part of this book may be reprinted, reproduced, transmitted, or utilized in any form by any electronic, mechanical, or other means, now known or hereafter invented, including photocopying, microfilming, and recording, or in any information storage or retrieval system, without written permission from the publishers.

For permission to photocopy or use material electronically from this work, please access www. copyright.com (http://www.copyright.com/) or contact the Copyright Clearance Center, Inc. (CCC), 222 Rosewood Drive, Danvers, MA 01923, 978-750-8400. CCC is a not-for-profit organiza-tion that provides licenses and registration for a variety of users. For organizations that have been granted a photocopy license by the CCC, a separate system of payment has been arranged.

Trademark Notice: Product or corporate names may be trademarks or registered trademarks, and are used only for identification and explanation without intent to infringe.

A Library of Congress record exists under LC control number: 83007093

Publisher's Note
The publisher has gone to great lengths to ensure the quality of this reprint but points out that some imperfections in the original copies may be apparent.

Disclaimer
The publisher has made every effort to trace copyright holders and welcomes correspondence from those they have been unable to contact.

ISBN 13: 978-1-138-50617-6 (hbk)
ISBN 13: 978-1-138-56166-3 (pbk)
ISBN 13: 978-0-203-71056-2 (ebk)

Visit the Taylor & Francis Web site at http://www.taylorandfrancis.com and the CRC Press Web site at http://www.crcpress.com

PREFACE

Recently, CRC Press published *Vitamins in Human Biology and Medicine,* a series of reviews summarizing modern aspects of the medical uses of various vitamins. The current volume is a continuation of that theme, presenting the results of recent research into vitamin biochemistry, physiology, and nutrition.

The book opens with a survey by Dr. Morton Cowan of biotin-responsive metabolic disorders of children. A group of Japanese researchers then present their latest data on the complex interactions between vitamin D and parathyroid hormone (PTH) in bone calcification. Dr. Maxine Briggs reviews the prolific published information relating vitamin C to infectious diseases and presents the results of an 8-year study into the prophylactic value of high-dose ascorbic acid (AA) and the common cold. Professor Kristoffersen and Dr. Rolschau review the effects of vitamin supplements during pregnancy and intrauterine growth. Possible protective effects of vitamin A and other retinoids against cancer are discussed by Dr. Jill Blunck, while Dr. Sue Tonkin describes her results on the interaction between oral contraceptives and riboflavin. Finally, the possible prevention of neural tube defects by vitamin supplements is described by Professor Laurence.

It is hoped that this collection of papers on the frontiers of vitamin research will be of wide interest to medical and scientific workers interested in the exciting and controversial uses and functions of vitamins.

Michael H. Briggs

THE EDITOR

Michael H. Briggs, D.Sc., Ph.D., F.R.S.C., F.I. Biol., F.R.C. Path., is Professor of Human Biology at Deakin University, Geelong, Victoria, Australia. Currently he is Dean of Sciences.

Born in Manchester, England in 1935, Dr. Briggs was educated at the University of Liverpool, then undertook postgraduate studies in biochemistry and nutrition at Cornell University and the University of New Zealand. He is a Fellow of the Royal Society of Chemistry, a Fellow of the Royal Institute of Biology, and a Fellow of the Royal College of Pathologists. Dr. Briggs is also a member of a wide range of national and international scientific and medical societies. Since 1970, he has been a Consultant to the World Health Organization, and in this capacity has visited 15 different countries.

Immediately prior to his present position, Dr. Briggs was Director of Biochemistry at the Alfred Hospital, Melbourne, a major teaching hospital of Monash University. In 1971 and 1972, Dr. Briggs was at the University of Zambia as Head of the Department of Biochemistry, where a Food and Agriculture Organization nutrition survey team formed part of his unit.

Dr. Briggs also spent 7 years in the pharmaceutical industry working as a research and development director on the therapeutic substances and nutrient additives for humans and farm animals.

Among his more than 200 publications are contributions to textbooks, monographs, and encyclopedias, as well as articles in scientific and medical periodicals on biochemistry, pathology, toxicology, pharmacology, enzymology, and reproductive biology. Dr. Briggs was editor of seven volumes of *Advances in Steroid Biochemistry and Pharmacology* and is co-editor of the new series *Progress in Hormone Pharmacology and Biochemistry*. Among other recent volumes he has written or edited are *Oral Contraceptives* (six volumes, 1977—1982, Eden Press), *Biochemical Contraception* (Academic Press, 1975), *Chemistry and Metabolism of Drugs and Toxins* (Heinemann Medical, 1974), *Pharmacological Models in Contraceptive Development* (WHO, 1974), *Implications of Steroid Hormones in Cancer* (Heinemann Medical, 1971), *Steroid Biochemistry and Pharmacology* (Academic Press, 1970), *Advances in the Study of the Prostate* (Heinemann Medical, 1970), and *Urea as a Protein Suppliment* (Pergamon Press, 1968).

Dr. Briggs is married to a medical practitioner, Dr. Maxine Briggs, who has been his research collaborator for many years and is co-author of many of their publications.

CONTRIBUTORS

Jill M. Blunck, M.B., B.S., Ph.D.
Senior Lecturer in Pathology and
  Toxicology
Deakin University
Geelong, Victoria, Australia

**Maxine Briggs, M.B., Ch.B., D.P.H.,
D.O.H., F.R.A.C.M.A.**
Director of Rehabilitation Services
Geelong Hospital
Victoria, Australia

Morton J. Cowan, M.D.
Assistant Professor of Pediatrics
University of California
San Francisco, California

Hiroyoshi Endo, Ph.D.
Professor of Physiological Chemistry
Faculty of Pharmaceutical Sciences
Teikyo University
Sagamiko, Kanagawa, Japan

Karl Kristoffersen, M.D.
Professor and Senior Chief of Obstetrics
  and Gynecology
Odense University Hospital
Odense, Denmark

Kohtaro Kawashima, Ph.D.
Associate Professor of Physiological
  Chemistry
Faculty of Pharmaceutical Sciences
Teikyo University
Sagamiko, Kanagawa, Japan

Mamoru Kiyoki, Ph.D.
Teijin Institute for Bio-Medical Research
Hino, Tokyo, Japan

**K. Michael Laurence, M.A., D.Sc.,
M.B., Ch.B., F.R.C.P., F.R.C. Path.**
Professor of Pediatric Research
Honorary Consultant Clinical Geneticist
Welsh National School of Medicine
University of Wales
Cardiff, Wales

John Rolschau, M.D.
Lecturer and Senior Registrar
Obstetrics and Gynecology
Odense University Hospital
Odense, Denmark

**Suzanne Y. Tonkin, B.Sc.(Hons),
M.Sc., Ph.D.**
Biochemist
Deakin University
Geelong, Victoria, Australia

TABLE OF CONTENTS

Chapter 1

BIOTIN-RESPONSIVE METABOLIC DISORDERS IN EARLY CHILDHOOD

Morton J. Cowan

TABLE OF CONTENTS

I. INTRODUCTION

Biotin, one of the B vitamins, is a cofactor for at least four carboxylation reactions in mammalian cells.[1,2] While it was known as early as the beginning of this century that raw egg white, which contains a biotin-binding protein (avidin), was a toxic nutrient, it was not until 1936 that biotin was characterized as a yeast growth factor.[3] It was subsequently shown to be identical to a variety of substances including "protective factor X," BIOS II B, vitamin H, and egg white injury factor.

Until relatively recently, interest in biotin focused on defining its role as a cofactor for supporting growth of microorganisms and understanding its metabolism and mode of action in eukaryotic and prokaryotic cell systems. It has only been in the last 20 years that the specific role of biotin as a cofactor for CO_2 transfer (carboxylation) processes has been elucidated.

The clinical significance of biotin-deficient and biotin-dependent states was not recognized until the early 1970s. Prior to that time it was felt that the GI flora produced sufficient biotin to prevent a deficient state unless a diet containing raw egg whites (avidin) was consumed.[4] A few case reports in the literature supported this assessment.[5,6] However, in 1971 the first inborn error of biotin metabolism was reported.[7] Subsequently, there have been numerous discoveries of children with either single or multiple deficiencies of biotin-dependent enzymes. In addition to the obvious effects of biotin deficiency on the skin and hair, it has become evident that in these affected children there are major disturbances of carbohydrate and amino and fatty acid metabolism involving not only acid-base and glucose homeostasis but CNS and immune system functions as well.

The current focus of biotin research remains on its metabolism but its role in humans has acquired a new significance since more biotin-deficient and biotin-dependent disorders have been discovered. The early diagnosis of patients with either acquired or inborn deficiencies in biotin metabolism has resulted in a simple and safe therapy for correcting one of the causes of organic aciduria in childhood. The deleterious metabolic, CNS, and immunologic consequences often can be reversed with pharmacologic doses of biotin and therapy may even begin during gestation following prenatal diagnosis. These biotin-responsive disorders clearly are models which can be used in approaching other vitamin-dependent states. Also, these new inborn errors have generated interest in the role of biotin-dependent metabolism in the immune system and CNS which may result in a better understanding of how biotin-dependent carboxylation selectively affects immune regulation as well as how it is involved in brain cell chemistry. Finally, by focusing attention on these pathways, a better understanding of the regulation of carbohydrate and amino and fatty acid metabolism will result and the potential always exists for the discovery of new pathways or metabolites, as yet unreported.

The subsequent sections will include: (1) a brief discussion of biotin chemistry and metabolism, (2) clinical and laboratory findings in children with biotin-responsive disorders, (3) the pathogenesis of these disorders, and finally (4) studies which have been done in selected areas on the effects of biotin deficiency.

II. BIOTIN CHEMISTRY AND METABOLISM

A. Chemistry

Biotin is a low molecular weight (M.W. = 244) coenzyme consisting of two fused rings, one of which is a ureido and the other a tetrahydrothiophene (Figure 1). The aliphatic side chain [$(-CH_2-)_4$ COOH] is essential for the covalent binding of biotin via the carboxyl group to an epsilon amino group of a lysine residue in the apoenzyme (Figure 2). During the

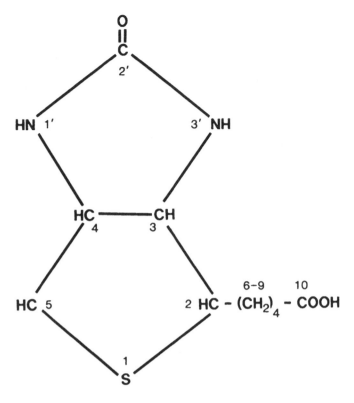

FIGURE 1.   Biotin.

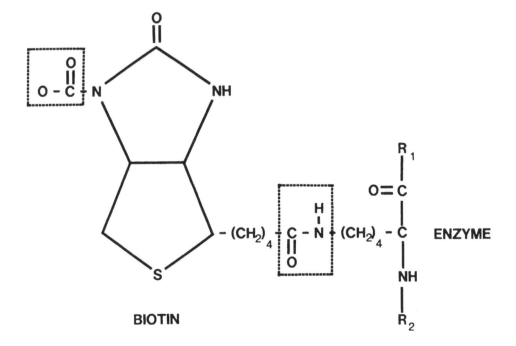

FIGURE 2.   Biotin binding to lysine residue of apocarboxylase.

Table 1
BIOTIN-DEPENDENT ENZYMES

Carboxylases
    Pyruvate carboxylase
    Propionyl-CoA carboxylase
    β-methylcrotonyl-CoA carboxylase
    Acetyl-CoA carboxylase
    Geranyl-CoA carboxylase
    Amino carboxylase
Transcarboxylases
    Methylmalonyl-CoA: pyruvate Carboxyltransferase
Decarboxylases
    Methylmalonyl-CoA decarboxylase
    Oxaloacetate decarboxylase

carboxyl transfer process, the CO_2 is carried by the N-1 nitrogen of the ureido ring. The active form of biotin is d-biotin. While it is soluble in water, its solubility significantly increases in dilute alkali and alcohol. Details of the chemistry of biotin including its isolation, structure, and chemical and physical properties have been reviewed.[1-8]

B. Biotin-Dependent Enzymes

There are nine known biotin-dependent enzymes[1] which can be divided into carboxylases, transcarboxylases, and decarboxylases (Table 1). Of these, four have been found in mammalian cells: pyruvate carboxylase, propionyl-CoA carboxylase, acetyl-CoA carboxylase, and β-methylcrotonyl-CoA carboxylase. The others have been described in various microbial organisms.

Another carboxylase (5-aminoimidazole ribonucleotide carboxylase) found in mammalian cells catalyzes in integral step in *de novo* purine biosynthesis. There are data in rodents suggesting that *de novo* purine biosynthesis is biotin dependent.[9] However, we have been unable to show biotin dependency of this carboxylase in other mammalian systems (unpublished data). Finally, carbamyl phosphate synthetase, important in pyrimidine and arginine biosynthesis also has been shown in *Escherichia coli* to be biotin dependent.[10] Early studies in biotin-deficient rats suggested that this enzyme may be biotin dependent in mammalian cells.[11] Confusion as to whether it is the mitochondrial (urea cycle) or cytosolic (*de novo* pyrimidine) enzyme which is affected (if at all) still exists and more recent evidence to support the biotin dependence of this enzyme is lacking. The remaining discussion will focus on the four known biotin-dependent carboxylases found in mammalian cells.

C. Biotin-Dependent CO₂ Fixation in Mammalian Cells

The four mammalian biotin-dependent enzymes affect carbohydrate and amino and fatty acid metabolism. It has been known since 1960[12] that pyruvate carboxylase (PC) is a biotin-dependent mitochondrial enzyme which metabolizes pyruvate, a central intermediate in carbohydrate metabolism (Figure 3). Under aerobic conditions the principal fate of pyruvate is conversion to acetyl-CoA by pyruvate dehydrogenase (PD). Another important reaction involving pyruvate utilizes PC which catalyzes a critical step in both gluconeogenesis and oxidative phosphorylation (tricarboxylic acid, TCA, cycle). PC catalyzes the carboxylation of pyruvate to oxaloacetate. This reaction is the initial step in gluconeogenesis in which oxaloacetate can then be converted to phosphoenolpyruvate by PEP carboxykinase. It is also an important means of replenishing an intermediate (oxaloacetate) of the TCA cycle. Acetyl-CoA is an allosteric activator of PC and therefore along with other substances (e.g., hormones) plays a critical role in regulating the supply of oxaloacetate for either gluconeogenesis or the TCA cycle. It is clear that the metabolic fate of pyruvate is carefully regulated and

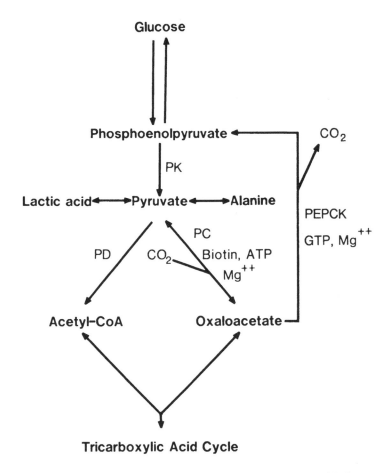

FIGURE 3. Major pathways of pyruvate metabolism: PD = pyruvate dehydrogenase; PC = pyruvate carboxylase; PEPCK = phosphoenolpyruvate carboxykinase; and PK = pyruvate kinase.

varies with the specific needs of different tissues and under a variety of different conditions. An abnormality in any of the major enzymes which metabolize pyruvate is likely to upset these delicate balances. In fact, the consequences of the absence of PC include decreased oxaloacetate, abnormal oxidative phosphorylation and gluconeogenesis, and increased pyruvate and lactate.

Two other biotin-dependent enzymes, located in the mitochondria, are propionyl-CoA carboxylase (PCC) and β-methylcrotonyl-CoA carboxylase (MCC). The former is involved in isoleucine and valine catabolism and the latter is necessary for leucine degradation. Leucine is catabolized to acetyl-CoA and acetoacetate, and therefore is a purely ketogenic amino acid. Isoleucine and valine are converted via methylmalonyl-CoA to succinyl-CoA, an important source of carbons for the TCA cycle. PCC also is involved in the catabolic pathway of methionine and threonine and participates in the oxidation of odd-chain fatty acids.

Finally, the fourth mammalian biotin-dependent enzyme is acetyl-CoA carboxylase (ACC), which initiates the synthesis of malonyl-CoA from acetyl-CoA. ACC is located in the cytosol and catalyzes this irreversible rate-limiting step in *de novo* synthesis of fatty acids which serve as a major source of energy storage as well as being components of complex lipids. It is interesting that acetyl-CoA is a mitochondrial precursor of citrate in the TCA cycle, and a major allosteric activator of ACC is citrate. The exact relationships between mitochondrial and cytosolic acetyl-CoA and citrate have not been well defined.

D. Biotin Synthesis, Absorption, and Transport

The biosynthesis of biotin has been studied thoroughly in microbial organisms such as *E. coli* and achromobacter.[8] While it is clear that normal (nonmalignant) mammalian cells do not synthesize biotin *de novo* the sources of biotin for mammalian metabolism have not been well defined, and probably vary among species.[8] The two major sources for man are diet and GI bacterial flora. A large number of foods have been found to contain varying amounts of biotin, with the best sources being egg yolk, soy flour, brewer's yeast, and liver. Even these foods, however, have relatively low amounts of biotin (100 to 200 μg/100 g). The fact that it is difficult to create an animal model of biotin deficiency unless the gut flora is altered with antibiotics or the biotin-binding glycoprotein avidin is added to the diet suggests that biotin synthesis by bacteria may be another important source of this vitamin. However, there are only a few detailed studies of intestinal biotin absorption which define where maximal absorption occurs and no detailed studies which define the relative significance of microbial synthesis and diet as sources of biotin.

Biotin metabolism in mammals consists of a complex series of steps including GI absorption, intravascular transport, cellular transport, intracellular transport, and catabolism. It appears that at least in the small intestine of the hamster and mouse (but not rat, rabbit, or guinea pig) biotin transport is carrier mediated against a concentration gradient[13,14] rather than absorbed by diffusion.[15] The portion of the small intestine in which maximal absorption occurs is in the proximal half closer to the pylorus than the ileocecal valve. Furthermore, this active transport process is sodium dependent, and the Km is 1.0 mM.

In man, it appears that the majority of intravascular biotin is bound to proteins, in particular the α- and β-globulins and albumin.[16] Detailed studies of serum biotin half-life and clearance have not been reported. One study of a single oral dose of biotin suggested a plasma half-life of about 3 hr.[17] Urinary clearances were not reported. Another study,[6] which was more detailed measured a whole blood half-life of about 3 hr. There are some early studies of biotin excretion in urine and stool indicating that total excretion in man is greater than oral intake while urinary output alone is usually less than intake.[4,18-20] In one experiment when the dietary biotin level was increased a small or moderate amount, urinary biotin excretion remained constant while fecal excretion increased significantly.[4] With a large increase in dietary biotin, urine excretion increased markedly relative to fecal output. The regulatory role of biotin absorption and/or renal clearance is difficult to ascertain from this study since the synthesis by gut flora was undefined. A major difficulty in all of the studies of biotin transport and clearance is the assay system used. It is apparent that different assays measure to varying degrees free and/or bound biotin. Until this variability is better defined interpretation of many of these results must be done cautiously.

The tissue distribution following an injection of labeled biotin provides some information about its metabolism. In one study[21] following intraperitoneal injection in rats, over 95% of the biotin was recovered in urine within 24 hr. The urinary excretion rate did not change in the biotin-deficient state.[22] In another study using chicks and rats it was found that 15 to 30% of intramuscularly injected biotin (labeled as $-C^{14}OOH$) was recovered in the liver and kidney while 30% was excreted.[23] Virtually no activity was found in free CO_2 suggesting little catabolism. It is difficult to compare these results since the biotin was administered in distinctly different ways. Finally, in a study[24] using dogs, the placental transfer of biotin was examined and found to be uninhibited. It was also found that prenatal and postnatal tissue biotin content (absolute levels per gram of tissue and H^3-biotin distribution) was greater than in adult animals, suggesting a change in biotin needs as a function of maturation.

Biotin transport into cells has not been well studied. There is some evidence from a malignant mammalian cell line that biotin transport occurs by simple diffusion without a carrier molecule.[25] This is in contrast to what has been found in some studies of intestinal absorption. Further work needs to be done to determine the validity of the findings. As far as the intracellular distribution of biotin, in one report[26] it appeared that about one third of labeled biotin was found in mitochondria with two thirds being found in the cytosolic fraction, and virtually no labeled biotin found in the nuclear fraction. However, in another report[23] 23% of the biotin was found in liver nuclei. While both studies used rats, the route of administration, dose of biotin, and extraction procedures differed significantly and could explain the varied results.

E. Biotin Binding to Apoenzyme and Carboxylation Processes

In the late 1950s it became clear that biotin was an integral part of a number of carboxylation reactions. The final step in the synthesis of carboxylase enzymes is one in which the carboxyl moiety of biotin is covalently bound to the apocarboxylase via the epsilon amino group of a lysine residue. This peptide linkage is catalyzed by holocarboxylase synthetase,[27,28] and is a two-step reaction which requires ATP and Mg^{++}. Holocarboxylase synthetase activity has been measured in mammalian cell extracts and has been found to have a Km for ATP of 0.2 mM and a Km for biotin of 8 nM.[29]

It is now clear that the carboxylation reaction catalyzed by biotin-dependent enzymes occurs in at least two intermediate reactions.[2] In the initial step carboxybiotin is formed from bicarbonate and requires 1 mol of ATP (Figure 4). The carboxyl group is attached to the 1'-nitrogen of the biotin ring (Figure 1). All of the carboxylase enzymes which have been studied have been found to be comprised of from 4 to 30 polypeptides, although the number of unique polypeptides is much less than 30. This structure varies significantly among species. It appears that there are separate subsites on the active enzymes for each partial intermediate reaction. Following the formation of carboxybiotin both the carboxyl carbon atom and the acceptor molecule must be activated. The exact sequence and mechanism by which the carboxyl group is transferred to the acceptor molecule is not completely known and at least two mechanisms have been proposed, one involving O-phosphobiotin formation[30] and the other involving a "concerted" mechanism.[31] In either case it appears likely that a two-step reaction involving multiple subsites of the biotin enzyme is involved.

F. Biotin Catabolism and Excretion

Very little is known about the fate of biotin during enzyme turnover. In a study using ^{14}C-biotin injected into rats the intact molecule was almost completely (95%) excreted in the urine within 24 hr with a small amount of the label in four metabolites.[21] When the rats were biotin deficient, there was no change in the rate of excretion.[22] These investigators, however, could find no evidence for extensive cleavage of the ring structure in contrast to what is found in bacteria.[32]

III. CLINICAL AND LABORATORY MANIFESTATIONS OF BIOTIN-DEFICIENT AND BIOTIN-DEPENDENT STATES

A. Biotin Deficiency

Until recently there had been only four reports of the effects of biotin deficiency on humans.[5,6,33,34] In one study,[33] four healthy adult volunteers were followed on a biotin-deficient diet containing raw egg white for approximately 2 months. Major organ systems which were involved included the skin, the nervous system, and the cardiovascular system. A maculosquamous dermatitis and atrophic glossitis developed in some of the subjects while nervous system symptoms occurred in all including changes in mental state, depression, hyperesthesia, paresthesia, and anorexia. There were also complaints of precordial chest pain and electrocardiographic evidence of cardiac ischemia. Laboratory abnormalities included anemia, hypercholesterolemia, and hyperbilirubinemia. Within a few days of biotin therapy the symptoms were resolved. Unfortunately, there were no controls who received

$$Mg^{++}$$

$$\text{STEP 1: E-BIOTIN + ATP + HO-CO}_2^- \rightleftharpoons \text{E-BIOTIN-CO}_2^- \text{+ ADP + Pi}$$

$$\text{STEP 2: E-BIOTIN-CO}_2^- \text{+ SUBSTRATE} \rightleftharpoons \text{PRODUCT-CO}_2 \text{+ E-BIOTIN}$$

FIGURE 4. Carboxylation reaction catalyzed by biotin-dependent enzymes.

the diet along with biotin supplementation and it is not clear as to what the contribution was of anorexia and decreased caloric intake to the findings.

The three additional reports were of individual patients whose diet included raw egg whites for one reason or another. One woman's course was complicated by cirrhosis and while she developed many of the symptoms reported in the study of normal subjects she did not develop abnormal electrocardiographic changes, hypercholesterolemia, or anemia.[6] Clearly her diet was significantly different from that of the earlier study and may explain the different findings. Another patient was a 5-year-old boy whose tube-feeding included 6 raw eggs daily.[5] This child developed a fine scaly dermatitis and alopecia. There were no nervous system or cardiac manifestations and the single abnormal laboratory finding was a significantly elevated serum cholesterol. The clinical and laboratory abnormalities rapidly corrected with biotin therapy. Additional studies suggested that the hypercholesterolemia may not have been due primarily to the biotin deficiency but rather the high egg white intake.

In 1979, Sweetman et al.[35] thoroughly evaluated an 11-year-old boy who presented with a 2-year history of alopecia totalis and an erythematous exfoliative dermatitis. This was associated with a 2 to 5 year history of a daily ingestion of raw eggs. Laboratory studies included a normal hemoglobin, cholesterol, glucose, lactate, pyruvate, and electrolytes. However, ketonuria was present and the organic acid in the urine was compatible with deficiencies of multiple biotin-dependent carboxylases. Plasma biotin levels prior to therapy were about 50% of the lowest control value, and the urine biotin was 1/40 the control mean level. White cell carboxylase activities were 29% of normal and increased with biotin therapy. Interestingly, this child had a past history of candida associated with the rash and presented with otitis media and a staphylococcal infection (see Section V.A.).

As the awareness of and interest in biotin metabolism in human disease has increased, another cause for biotin deficiency has been found to be a potential complication of biotin-free parenteral alimentation.[36,37] In each instance the patients had significant GI pathology (short gut syndrome in one, and chronic severe secretory diarrhea in the other), necessitating total parenteral alimentation. In both instances, there was a periorificial candida dermatitis and alopecia. Organic aciduria and low serum and urine biotin levels were diagnostic of significant biotin deficiency and a rapid response to biotin supplementation was confirmatory.

Throughout the years attempts have been made to associate biotin deficiency with a variety of clinical states. For example, the skin manifestations have suggested a possible role of biotin deficiency in seborrheic dermatitis and Leiner's disease.[38] Leiner's disease was initially described as occurring in breast-fed infants 1 to 2 months after birth. It was diagnosed clinically based upon scalp crusting, an intense generalized erythema accompanied by whitish-gray scales most marked in the exterior surfaces, ridging of the nails, significant diarrhea, and failure to thrive. For many years Leiner's disease was thought to be a more severe form of seborrheic dermatitis of infancy and it was not until recently that it was found to be due to a deficiency of the fifth component (C5) of complement.[39]

However, there are several studies which describe infants with either Leiner's disease or seborrheic dermatitis who have responded to biotin therapy.[38,40,41] In some instances serum and urine biotin levels were shown to be abnormally low in affected patients. However, not every patient responded to biotin. Two recent double-blind controlled trials of biotin therapy in seborrheic dermatitis failed to demonstrate any significant effect of biotin compared to a placebo control[42,43] What is likely to account for these widely discrepant results is the fact that multiple etiologies probably exist for these skin disorders, at least two of which are a deficiency of C5 and biotin. This is comparable to the clinical state of alopecia found not only in biotin deficiency (or dependency), but in certain autoimmune and endocrine diseases as well as disorders in vitamin D metabolism.[44] Detailed biochemical studies are needed on patients with seborrheic dermatitis and/or Leiner's disease to try to sort out these issues more effectively.

Two other examples of attempts to associate biotin deficiency with clinical disorders involve children with burns and scalds[45] and sudden infant death syndrome (SIDS).[46] In both studies biotin content in serum or liver was found to be lower than normal controls. In neither report was there any measurements made of organic acid levels or carboxylase activities which would aid in interpreting the significance of the findings. Also, measurements of other vitamins were not reported for comparison. In children with burns there are many possible reasons for a low biotin level including nutritional support, antibiotic therapy, and secondary GI disease. Certainly, it is conceivable, but not known, that secondary biotin deficiency in critically ill children can worsen their course or prolong their recovery. The cause(s) of SIDS is not known and nutritional deficiencies have been suggested. One study[46] reported only children under 1 week of age (or stillborn) had liver biotin levels that were at the lower end of the 95% confidence limits for normal. The authors postulate that relatively low biotin levels increase the susceptibility of the infant to another stress resulting in SIDS. Their results neither confirm nor support this hypothesis. Further studies in both burn patients and populations susceptible to SIDS will have to be done to define the significance of biotin in either state.

B. Inborn Errors of Biotin-Dependent Enzymes

In 1966, the first application of gas-liquid chromatography and mass spectrometry to diagnose organic acidurias resulted in the discovery of isovaleric acidemia, an inborn error of leucine catabolism. Using similar technology Hommes et al.[47] in 1968 described the first inborn error involving a biotin-dependent enzyme, propionyl-CoA carboxylase. Since then patients have been reported with separate deficiencies of two other biotin-dependent enzymes, pyruvate carboxylase and β-methylcrotonyl-CoA carboxylase (Table 2). Recently, there has been a single report of a child with possible acetyl-CoA carboxylase deficiency.[48]

The first patient found to have a deficiency of all four biotin-dependent enzymes (multiple carboxylase deficiency, MCD) initially was reported (1971) as having β-methylcrotonyl-CoA carboxylase deficiency.[7] Additional studies[49,50] described in 1977 demonstrated MCD. Both this patient and a second one with MCD[51] reported in 1976 had similar clinical and biochemical characteristics and are now referred to as the early onset or neonatal form of MCD. In 1979, we reported[52] a third family with MCD in which there were distinct clinical and biochemical characteristics compatible with a later-onset or infantile type (Table 3). To date, evidence exists that both types of biotin-dependent carboxylase deficiency are inherited as autosomal recessive defects, that is, affected male and female children have been found in the same sibship. The following sections will cover the individual biotin-dependent carboxylase deficiencies and the inborn errors in biotin metabolism resulting in MCD.

1. Propionyl-CoA Carboxylase (PCC) Deficiency

At least several dozen patients have been found with propionic acidemia due to PCC

Table 2
INBORN ERRORS OF BIOTIN-DEPENDENT ENZYMES

Enzyme deficiency	Onset of symptoms	Clinical/lab immunodeficiency	Neurologic symptoms	Laboratory findings	Organic acids
PCC[a]	First year	+	Hypotonia, areflexia	Severe, persistent ketoacidosis, hyperammonemia	Propionic acid, methylcitrate
MCC	First year	+	Hypotonia, tongue fi-brillation, delayed motor	± Ketoacidosis	HIVA, MCG[b]
PC	Neonatal	?	Severe mo-tor/mental delay, seizures	Lactic acidosis, hy-poglycemia, hyperammonemia	Lactic acid
CC	Neonatal	?	Hypotonic myopathy, mental retardation	?	Derivatives of hex-anoic acid, acid, 2-ethyl hexanedioic acid

[a]   PCC = propionyl-CoA carboxylase; MCC = methyl-crotonyl-CoA carboxylase; PC = pyruvate carboxylase; ACC = acetyl-CoA carboxylase.

[b]   HIVA = hydroxyisovaleric acid; MCG = methylcrotonylglycine.

Table 3
CLINICAL MANIFESTATIONS OF MCD

	Early (neonatal)	Late (infantile)
Inheritance	Autosomal recessive	Autosomal recessive
Onset	Newborn—3 mos.	>3 mos.
Neurologic symptoms	Hypotonia seizures, lethargy, coma	Seizures, intermittent ataxia, hypotonia
Immunologic findings	Bacterial, viral in-fections, pneumonia	Candida dermatitis, bacterial, viral infec-tions, pneumonia
Skin manifestations	± Alopecia ± Dermatitis	Alopecia Periorifical dermatitis

deficiency.[53] These patients usually, but not always, develop symptoms within the first year of life. In some, while not acutely ill at birth, there is the development within a few weeks to months of life of a persistent severe ketoacidosis associated with vomiting, hyperventilation, grunting, and apnea. Neurologic symptoms consist of hypotonia, areflexia, and seizures. These patients have a very high morbidity and mortality. In others, the presentation may not be until after the first year of life in which there are recurrent episodes of ketoacidosis which may progress to a clinical picture which is similar to the early onset disorder. Survival may be better in this group if the diagnosis is recognized; if protein is restricted in the diet, normal physical and mental growth may be achieved.[54] There are even reports[55] of an affected child with symptomatic siblings who was totally asymptomatic. However, there is a significant incidence of mental retardation among those children experiencing severe episodes of ketoacidosis. In many of these children the episodes of severe acidosis are associated with or induced by infectious illness. In one child this was associated with evidence of immune dysfunction,[56] and in others there has been associated neutropenia, thrombocytopenia, or pancytopenia.

The laboratory findings in children with PCC deficiency usually include metabolic acidosis, hyperglycinemia, propionic acidemia, and hyperammonemia. However, there are not always consistent laboratory findings and patients have been reported with minimal to no acidosis in spite of severe illness,[45,57] without hyperglycinemia,[53,57] and with normal or only slightly elevated blood and urine propionic acid.[53] The mechanisms for the elevated ammonia and glycine in PCC deficiency are not known although it is possible that they reflect secondary abnormalities due to toxic effects on mitochondrial enzyme systems. One urine metabolite of propionic acid which has been found in the neonatal and late onset forms is methylcitrate which is of prenatal diagnostic value.[58]

PCC deficiency is inherited as an autosomal recessive defect. Genetic complementation studies indicate that at least two major groups (pccA and pccC) and two minor groups (pccB and pccBC) exist and in each there is evidence of the presence of an abnormal mutant enzyme.[53,59-61] The clinical types seen in PCC deficiency do not correlate with complementation group assignment. In both major groups there is evidence for single locus involvement.

Therapy for PCC deficiency has consisted of either dietary restriction or pharmacologic biotin administration. The relatively poor in vitro and in vivo response to biotin in many patients with this defect has made dietary protein restriction combined with careful monitoring of ketonuria the major approach. However, results from a recent study of biotin responsiveness in 15 patients with propionic acidemia suggest that biotin supplementation should be considered a part of the therapeutic regimen for all affected patients.[62] Finally, until more is known about the effects of propionic acidemia on immune function, and the findings of immunodeficiency in at least one patient, consideration should be given to avoiding exposure to live viral vaccines and varicella (see Section V.B.).

2. β-Methylcrotonyl-CoA Carboxylase (MCC) Deficiency

There have been only a few patients reported with possible MCC deficiency.[63,64] The clinical manifestations of this disorder are variable, but in those patients they may be fatal without therapy. One patient who presented in the second week of life had neuromuscular symptoms of dysphagia, delayed motor development, severe hypotonia, muscle atrophy, absent deep tendon reflexes, and fibrillation of the tongue while being mentally alert.[63] She did not have ketoacidosis but excreted large amounts of β-hydroxyisovaleric acid (HIVA) and β-methylcrotonylglycine (MCG) in her urine. A second child also presenting in the first year of life had severe ketoacidosis associated with an extensive dermatitis, but no neuromuscular disease. A third patient with MCC deficiency presented at 9 months of age with a history of recurrent respiratory infections and delayed motor development but without rash or ketoacidosis.[64] This child was moribund and in respiratory failure at the time of diagnosis with a severe pneumonia. Interestingly, the first patient described died with bronchopneumonia although this may have been a result of aspiration.

Diagnosis of MCC deficiency is suggested by the finding in urine of characteristic metabolites (HIVA and MCG), and confirmed by measuring absent MCC activity in fibroblasts or peripheral blood leukocytes. At least one of the metabolites (HIVA) is normally detected in low levels in the urine and may be nonspecifically elevated in experimental and clinical ketoacidosis not due to MCC deficiency per se.[65] Unfortunately, in at least two of the patients reported to have MCC deficiency[63,64] results of enzyme analyses were not mentioned.

Therapy for MCC deficiency is aimed at dietary restriction of leucine if there is no response to biotin. However, the one patient who did not respond to pharmacologic doses of biotin died in spite of protein restriction. Too few patients have been studied to define either the biochemical genetics or the clinical types.

3. Pyruvate Carboxylase (PC) Deficiency

Only a few patients have been reported with absent PC.[66,69] The onset of symptoms have often been in the neonatal period and have almost always included severe neurological impairment. These children have severe developmental delay and mental retardation, often with seizures. When PC activity has been extremely low or absent the outcome has been fatal. At least some children have had clinical and pathological findings compatible with subacute necrotizing encephalomyelopathy (Leigh's encephalopathy).[66,67] However, it is not at all clear to what extent PC deficiency is responsible for this disease or whether a low (but not absent) PC activity is a secondary manifestation of another primary defect or an artifact of poor sample preservation since numerous patients with Leigh's disease have been found to have normal PC activity.[70]

Laboratory findings include lactic acidosis, hypoglycemia, generalized aminoaciduria, hyperammonemia, and significantly elevated blood lactate and pyruvate with a high lactate:pyruvate ratio. At least one patient has been found to have proximal renal tubular acidosis.

Diagnosis should be considered in any child with a lactic acidosis and fasting hypoglycemia. It is confirmed by demonstrating absent or reduced PC activity in liver, fibroblasts, or blood leukocytes.[71] In those children with only a partial deficiency, i.e., 10 to 25% of normal, it is difficult to determine the exact role of PC deficiency in the disorder. Therapy has focused on efforts to maintain a normal glucose level and minimize lactic acidosis with frequent carbohydrate feedings. Too few patients have been reported to know the effectiveness of biotin therapy in this disease.

4. Acetyl-CoA Carboxylase (ACC) Deficiency

There has been a single report of a girl presenting in the newborn period with what appeared to be ACC deficiency.[48] This child had severe hypotonic myopathy and brain damage. Specific organic acids in her urine suggested a defect in *de novo* fatty acid synthesis. ACC activity was found to be very low (2%) in liver and low (10% of control) in cultured skin fibroblasts, while PCC activity was normal.

C. Inborn Errors of Biotin Metabolism

Since the first description of a child with multiple carboxylase deficiency (MCD) in 1971,[7] there have been at least 18 affected children from 12 families reported in the literature.[7,52,72-80] There are two general clinical types of MCD (Table 3). There is also biochemical evidence for a genetic heterogeneity in MCD which correlates to some extent with the clinical findings.[81] Both are inherited in an autosomal recessive pattern. The early neonatal onset type has been associated with a defect in holocarboxylase synthetase while the later onset infantile form of MCD is thought to reflect defective GI transport of biotin.[82,83]

The clinical presentations of six children with neonatal onset MCD have been reported in detail.[7,72-77] All of them developed symptoms in the first 3 months of life, three within 1 to 2 days after birth. Siblings with a postmortem diagnosis or history suggestive of MCD deficiency had presentations which were similar to the proband in the family. The majority (7/10) of children who presented in the newborn period had acute onset of neurologic disease characterized by hypotonia, seizures, and lethargy often progressing to coma. This was invariably associated with severe metabolic acidosis. In one case[73] several episodes of bacteremia with different pathogenic organisms were part of the clinical presentation. In another case,[74] a clinical and laboratory picture of diffuse intravascular coagulation was seen. This child also had congenital heart disease unrelated to the MCD. In three of the reported children with the neonatal form of MCD, the onset of symptoms was later (6 weeks to 3 months) and more gradual, and involved other manifestations of biotin deficiency including alopecia and an extensive erythematous, exfoliative skin rash. CNS symptoms included irritability, lethargy, hypotonia, and developmental delay. In two of these children[76,77] a life-threatening event did not occur prior to diagnosis while in one child[75] symptoms progressed to shock prior to therapy. In this case the child had been exposed prenatally to biotin because of a significant family history suggestive of MCD, but did not receive biotin

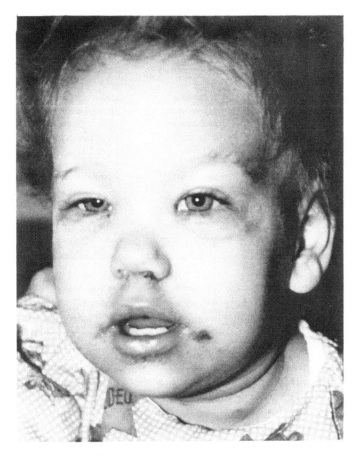

FIGURE 5. Child with late onset MCD.

during the first 3 months of life. The prior exposure to biotin therapy complicates interpretation of his clinical presentation. Another child[7] presented with persistent vomiting without failure to thrive, abnormal behavior, and the gradual onset of deep, rapid respirations. This child also had a rash which extended over the buttocks, neck, and flexure areas of the body. In several patients with neonatal onset MCD an unusual odor to the urine has been noted resembling "cat's urine".

Seven children in five families have been reported in sufficient detail to characterize the clinical manifestations of the late or infantile onset MCD.[52,78-80] These patients all presented between 3 months and 3 years of age. The development of symptoms generally was gradual and associated with episodes of remission in some but not all patients. Neurologic manifestations included grand mal and myoclonic seizures, intermittent truncal ataxia, and hypotonia. The skin disease was characterized by alopecia (in some instances, alopecia totalis), eczematoid dermatitis, periorificial candida dermatitis, and keratoconjunctivitis (Figure 5). Motor retardation and delayed development was seen in some but not all of these children. In at least two cases[52,79] the skin findings were consistent with acrodermatitis enteropathica although zinc levels were normal and in one case zinc therapy was ineffective. In addition to the candida dermatitis in four children, one of which was successfully treated with intravenous amphoteracin, there was a history of repeated bacterial and viral infections in at least one family.[52] Two children presenting at 3 and 6 months, respectively progressed to life-threatening episodes of ketoacidosis, dehydration, and coma.[80]

The diagnosis of MCD should be suspected in any child presenting either in the neonatal period or first few years of life with metabolic acidosis and the other clinical manifestations

Table 4
LABORATORY FINDINGS IN MCD

	Early (neonatal)	Late (infantile)
Plasma	Lactic acidosis	Lactic acidosis
	Hyperglycenemia	Hyperglycenemia
	Hyperammonemia	Hyperammonemia
	Hypoglycemia	Hyperphosphatemia
Plasma biotin	Normal	Low
Urine	Ketones	Ketones
	Organic acids[a]	Organic acids[a]
Carboxylase activity		
Cultured fibroblasts	Low to absent	Normal
Fresh leukocytes	Low to absent	Low to absent

[a]   Methylcitrate, propionic acid, hydroxyisovalerate, and methylcrotonylglycine.

of biotin deficiency, i.e., CNS disease (seizures, ataxia, hypotonia) and skin disease (dermatitis and/or alopecia) which may be associated with frequent viral, bacterial, or cutaneous fungal infections. Laboratory findings in both the neonatal and infantile forms include lactic acidosis with increased blood lactate and pyruvate and ketonuria (Table 4). A nonspecific amino aciduria and hyperglycinemia has also been found in many patients, as well as slight to moderate hyperammonemia. Organic acids in the urine are metabolic products of the biotin dependent pathways involving pyruvate, branched chain amino acid and *de novo* fatty acid metabolism. It should be noted that the ketoacidosis is intermittent in some patients although in all of them the organic aciduria is a consistent and diagnostic finding. Enzyme assays of leukocyte carboxylase activities in untreated patients confirm the diagnosis.

There are some laboratory findings which may distinguish children with neonatal and infantile onset MCD. One major laboratory finding in the neonatal form that is different from the infantile MCD is low to absent carboxylase activities in skin fibroblasts cultured in media which is biotin depleted.[81] The enzyme activities in infantile MCD fibroblasts and control fibroblasts, both grown in biotin-depleted media, are indistinguishable.[82] This finding has significant diagnostic implications, in particular with respect to prenatal diagnosis of MCD. Another laboratory finding in the infantile MCD which distinguishes it from the neonatal type is significantly low serum and urine biotin levels.[52,78] This has been used to support the hypothesis of a defect in biotin transport as the cause of the infantile onset disorder. There are several other differences between the two types of MCD which are less distinct. The neonatal or early onset form has been associated with hypoglycemia and moderately elevated blood ammonia levels. At least one child with the late onset form had a slightly elevated ammonia[52] so that this in fact may not be a distinguishing feature. In addition, at least one family with three affected children with late onset disease had consistently elevated phosphates with normal calcium and parathormone levels.[52] Hyperphosphatemia has not been mentioned in other case reports so the significance of its presence (or absence) cannot be determined although it may represent a secondary effect of acidosis on renal tubular function. In fact hyperphosphatemia has been associated with lactic acidosis.[84] The elevated blood ammonia and hyperglycinemia are also thought to reflect effects on other metabolic pathways secondary to the MCD.[53,85]

Therapy for all patients with MCD includes pharmacologic doses of biotin. To date, there have been no patients reported to be entirely resistant to biotin therapy, although at least one patient has shown only a partial response.[72] Both intravenous and oral preparations have been used successfully in varying dosage schedules. As little as 1 mg of biotin daily to as much as 40 mg a day has been given. There are no known adverse effects of biotin in either

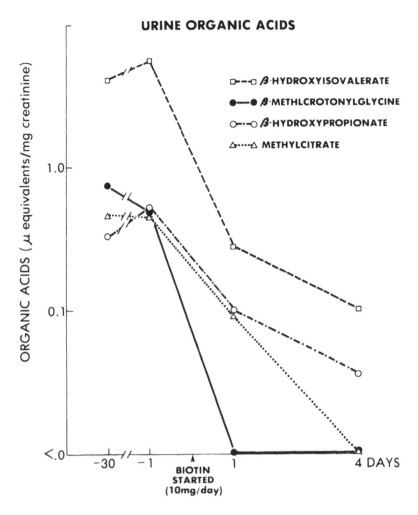

FIGURE 6. Reponse of organic aciduria in MCD to biotin therapy.

animal studies[8] or in patients. The optimal dose as well as the dosing schedule have not been determined. In one patient, when only a standard multivitamin preparation containing small amounts of biotin was given, there was no improvement in the clinical status; in fact, the patient ultimately died. This further emphasizes the fact that MCD is a biotin-dependent state and as such is distinct from acquired biotin deficiencies. In addition to biotin, other therapeutic considerations may involve frequent carbohydrate feedings and fatty acid supplementation. Finally, because of the increased susceptibility to infection we feel these children should not receive live viral vaccines, and febrile illnesses should be managed carefully.

The response to biotin in all patients with MCD is dramatic. Within hours there is significant clinical improvement, especially in those patients who have progressed to coma. In all patients with dermatitis and alopecia, skin clearing and complete hair restoration have occurred within a few months of therapy. The neurologic manifestations have also responded to therapy and in most instances have appeared to be completely reversible. At least one patient with MCD has demonstrated subtle yet measurable defects in detailed psychological testing in spite of therapy. Significant normalization in laboratory findings including organic aciduria (Figure 6), ketoacidosis, hyperammonemia, and hypoglycemia occurs within hours to days following biotin administration.

Neonatal onset MCD has successfully been diagnosed in utero using cultured amniotic fluid fibroblasts.[86] Prenatal biotin therapy has been given to two affected infants, both of whom were clinically normal at birth. In one, the diagnosis was made in utero,[86] and in the other it was made postnatally after the infant developed symptoms while off biotin therapy for 3 months.[75] In both, 10 mg of biotin was given orally to the mother during the third trimester and in neither was there evidence for a teratogenic effect. There are currently no methods available for the prenatal diagnosis of the infantile onset type of MCD.

IV. PATHOGENESIS OF BIOTIN-DEFICIENT AND BIOTIN-DEPENDENT STATES

There are at least two defects in the metabolism of biotin which result in MCD.[81] They were suspected when we found that the cultured fibroblast carboxylase activities were normal from a patient who clinically and biochemically had MCD.[82] This was in contrast to studies in fibroblasts from the initial patient reported.[87] We also noted low serum and urine biotin levels in one patient, a finding which was different from that noted in some of the other patients and which has since been confirmed in patients with infantile onset disease.[52,78]

Recently, two laboratories have measured the activity of holocarboxylase synthetase (HCS), an enzyme which catalyzes in the presence of ATP, the covalent attachment of biotin to apocarboxylase.[29,76] In fibroblasts or lymphocytes from two patients with the neonatal form of MCD the Kms for biotin in the HCS assay were 693 ng/mℓ and 648 ng/mℓ (normal = 2.0 ng/mℓ and 1.0 ng/mℓ, respectively). Vmax for each patient was 30% of normal. In our patient with neonatal MCD, a similar defect has been found.[86] These results suggest that in these patients a mutation has occurred in the HCS resulting in a functionally abnormal enzyme. The fact that both cytosolic (PC, PCC, MCC) and mitochondrial (ACC) carboxylases are effected indicates that the same HCS is responsible for apocarboxylase activation with biotin. Complementation studies using cultured skin fibroblasts from two patients with neonatal MCD and 14 patients with PCC deficiency support the concept of a mutant HCS.[88] The MCD fibroblasts do not complement one another although they complement all of the PCC deficient fibroblasts which mapped to pccA, pccBC, pccB, or pccC. Further studies with other MCD cell lines are needed to determine whether other mutants exist.

Results of studies in patients with infantile onset MCD suggest that the defect in biotin metabolism resulting in MCD is prior to the intracellular utilization of biotin. Normal carboxylase activities in cultured skin fibroblasts support this conclusion. At what step(s) in this pathway the defect occurs is not known. However, depressed urine and serum biotin levels suggest an abnormal intestinal or plasma transport of biotin. In addition, evidence for delayed absorption of biotin in one patient supports an abnormal GI transport.[89] However, while biotin levels are low in these children, they are not absent and one wonders whether the defect might also affect transcellular transport in tissues other than intestinal mucosa. This is supported by the finding of absent or very low carboxylase activities in peripheral blood leukocytes from these patients prior to therapy. With a Km for biotin of 1 to 2 ng/mℓ a biotin level which was even 50% of normal should result in no more than approximately a 50% decline in the velocity of reaction. Intracellular levels of biotin have not been measured although the enzyme results in cultured fibroblasts (i.e., they behave as normal controls in biotin deficient media) indicate a significant cellular transport defect does not exist, at least in fibroblasts. Detailed pharmacokinetic studies of biotin absorption in normals and in children with infantile MCD have not been performed. A better understanding of the GI absorption of biotin as well as cellular transport is needed in order to define the specific defect in these children.

These two defects in biotin metabolism can explain to some extent the different clinical manifestations found in MCD. Children with a defective HCS would be expected to be more sensitive to stress and in fact, should have carboxylase abnormalities in utero and at birth.

This would certainly explain an acute early onset of symptoms. Those children with normal HCS but with defective transport presumably can get some biotin into the cell and, while at low levels, should have some functioning carboxylases. Therefore, during fetal life, biotin can be transported across the placenta and these children would be expected to be normal at birth. The development of symptoms of biotin deficiency would be more gradual and might be intermittent depending upon diet as well as stresses such as febrile illnesses or fasting. In either type of defect the degree of symptoms appears related to a number of known (and still unknown) factors such as the extent of biotin dependency (i.e., association constant or Km of mutant protein), type of external stress, and the variations in abilities of other pathways to compensate for the defective carboxylases. Both types are fatal if untreated.

In spite of what is known about these two abnormalities in biotin metabolism there is significant clinical heterogeneity which has not been explained. For example, at least one child with the early onset (neonatal) form of MCD presented at 4 years of age with a history of only two episodes of metabolic acidosis.[77] This patient had low carboxylase activities in skin fibroblasts cultured in deficient media. It is not known why her clinical picture was more like the late onset form of MCD. Another child with documented HCS deficiency had a more gradual clinical course until 5 months of age, unlike the acute neonatal type.[7] Finally, at least several children with the biochemical phenotype of the infantile onset MCD have had acute episodes more compatible with the early onset picture although none have presented in the newborn period.[80]

Some of the clinical and laboratory findings can be explained by the deficient carboxylase enzymes. The skin manifestations are at least in part due to ACC deficiency resulting in defective *de novo* fatty acid metabolism.[90] The dermatitis has responded in some children treated with long chain fatty acids although the alopecia has persisted. The hypoglycemia is most certainly due to defective gluconeogenesis secondary to low PC activity. The role of PCC and MCC deficiency in the clinical and laboratory manifestation of MCD are not as clear. The organic acids which build up are probably in large part due to these deficient enzymes and it is thought that these organic acids are toxic to other essential metabolic processes. In addition, the catabolic products of branched chain amino acids are both glucogenic (succinyl-CoA) and ketonogenic (acetyl-CoA and acetoacetic acid), and there is sufficient evidence to suggest that a defect in catabolism of these amino acids significantly affects a number of metabolic pathways including protein synthesis and degradation, gluconeogenesis, lipid synthesis, and energy metabolism.[91] Finally, the neurologic manifestations and the effects on the immune system are not understood and will be reviewed in more detail below.

V. SELECTED TOPICS

A. Neurologic Diseases and Biotin Metabolism

It is clear that there are major effects of biotin deficiency on the nervous system. These have been recognized for a long time in several animal species including man.[8] In addition, children with both neonatal and infantile forms of MCD have symptoms reflecting CNS disease having both acute and chronic implications. The symptoms vary among patients; however, hypotonia, irritability, and seizure activity appear to be quite frequent and common findings. In addition, children with acute disease (neonatal) often have a history of progressing to severe lethargy and coma prior to diagnosis and therapy, while those with a more chronic form (infantile) develop intermittent ataxia and developmental delay. However, in two children with documented HCS deficiency who had a more chronic type clinical picture, developmental delay was a feature of their disease.[7,76] Also, several children with a putative GI transport defect had episodes of coma as part of their clinical presentation.[80] It is likely that if children with the acute neonatal form survived long enough without specific therapy, more would have a history of retardation and ataxia. Conversely, if those patients with a putative transport defect had a severe stress or were unable to compensate adequately, they likely would have CNS symptoms of a more acute nature.

There is at least one report of detailed neuropathologic examination of brain tissue from a child who died with MCD prior to diagnosis.[92] This child was one of three affected children, two of whom died prior to diagnosis.[52] His clinical course was compatible with the infantile, chronic type of MCD. He began having myoclonic and generalized clonic seizures at 3 months of age which were unresponsive to therapy. Intermittent ataxia was noted at 21 months of age. The patient received a live mumps vaccine 6 weeks prior to his death. Following the immunization he deteriorated neurologically and died at approximately $3^{1}/_{2}$ years of age. Microscopic examination of his formalin-fixed brain revealed evidence for a chronic cerebellar degeneration, a subacute myelopathy, and an acute meningoencephalitis. The chronic cerebellar disease was characterized by virtually complete loss of the Purkinje cell layer, rarefaction of the granular layer, and marked proliferation of the Bergmann layer. These findings were most severe in the superior vermis and adjacent mesial parts of the hemispheres. These chronic changes were most likely the direct result of the MCD. The subacute necrotizing myelopathy affected focal parts of the posterior, lateral, and anterior columns of white matter and was most compatible with a mumps infection. Immunofluorescent staining and viral cultures were not done. The acute meningoencephalitis also probably had a viral etiology although none was proved. Other detailed studies of the neuropathology in MCD have not been reported.

There are other inborn errors of metabolism which are associated with some neurologic features of MCD, in particular, intermittent ataxia. They include Hartnup disease, intermittent branched chain ketonuria, pyruvate decarboxylase deficiency, acrodermatitis enteropathica, and ornithine transcarboxylase deficiency. At least some of these disorders involve pyruvate or branched chain amino acid metabolism. There are a number of other inborn errors in which ataxia is also a feature including one associated with immunodeficiency (ataxia-telangiectasia) however, the other features of these diseases are quite distinct from MCD.[93] One exception is maple syrup urine disease which is associated with mental retardation, irritability, ataxia, and seizures. This inborn error is due to a deficiency of branched chain keto acid decarboxylase, a step which is part of branched chain amino acid catabolism. The relationship between the CNS findings in this disorder and in MCD is not known. Another probable inborn error of metabolism associated with lactic acidosis and ataxia which is thought to involve pyruvate metabolism is Leigh's subacute necrotizing encephalitis. At one time pyruvate carboxylase deficiency was postulated to be the cause of this disorder, however, current thinking has focused on the pyruvate dehydrogenase pathway.[94] In any case, patients with Leigh's encephalopathy are characterized by somnolence, blindness, deafness, and spasticity, none of which are features of MCD. Furthermore, the pathologic findings in this disorder are distinctly different from those seen in MCD.

It is interesting that many of the chronic neuropathologic changes seen in our patients with MCD[92] resemble what has been found in patients with chronic alcoholism, in particular the restriction of changes to the superior vermis and adjacent areas of the cerebellar hemispheres. The cerebellum is known to be sensitive to agents or conditions such as alcohols, anoxia, and probably the metabolites found in some inborn errors. It has been postulated, though not proven, that thiamin deficiency in chronic alcoholism results in a lactic acidosis secondary to low pyruvate decarboxylase, an enzyme which is dependent on thiamin pyrophosphate for normal activity.

The findings of CNS disease associated with disorders of pyruvate metabolism and lactic acidosis, have resulted in speculation that there may be a differential distribution of pyruvate carboxylase (PC) throughout the brain making certain areas more susceptible to PC deficiency than others.[94] At least two studies have looked at the distribution of PC in the brain in rats on a biotin-deficient diet.[95,96] In one study,[95] the effect of biotin deficiency on developing rat brains was examined by exposing pregnant rats and their offspring to a biotin-deficient diet. Whole brain PC activity in the deficient weanling animals was 70% lower than in the controls while liver activity was 85% lower. When different parts of the normal rat were examined, the highest PC activity was in the brain stem compared to cerebrum and cerebellum. The relative effect of biotin deficiency on PC activity in different parts of the brain was about equal. While other biotin-dependent enzymes were not measured in this study, it is interesting that metabolites of branched chain amino acid metabolism which should have been elevated did not suggest that PCC and MCC were not significantly lowered in this model. Also, light and electron microscopy of the brains from these animals showed no pathologic changes, raising the possibility that the deficiency which was induced was inadequate.

The other study[96] looked at rats fed a biotin-deficient diet at the time of weaning. PC activity was measured in freshly prepared mitochondrial pellets. In the normal adult rats no differences could be found in PC activity between gray and white matter. Also, the activities in the cerebellar hemispheres and vermis and in the cerebrum and brain stem were similar. Activity in the cerebellum (vermis or hemispheres) was highest and significantly greater than that found in the cerebrum or brain stem. In the deficient animals, hepatic PC activity was 3% of normal indicating significant biotin deficiency. In contrast PC activity in the different parts of the brain ranged from 53 to 71% of control. As in the other study, histologic examination revealed no differences from control. These studies suggest that selective PC deficiency cannot account for the neurologic manifestations in MCD, assuming that the biotin-deficient rat is an adequate model for MCD. It is interesting that the brain PC activity is in fact preserved relative to the liver and raises some as yet unanswered questions as to the mechanism and purpose for a relative sparing of the brain pyruvate metabolism in biotin deficiency. These findings also suggest the possibility that the biotin-deficient rat may not be a good model of MCD. Further studies will be necessary to explore these issues.

B. Immunodeficiency and Biotin Metabolism

The first patient in whom we diagnosed infantile MCD was referred to us for evaluation for mucocutaneous candidiasis.[52] One of her brothers had candida dermatitis, IgA deficiency, and absent antibody reponses to polysaccharide and protein antigens. In addition, following a live mumps vaccination he deteriorated and died 6 weeks later. Pathologic examination of his brain revealed evidence of mumps encephalitis, supporting the likelihood of a cellular immune defect.[92] Studies of T cell immunity in both children revealed absent lymphocyte responses to candida antigen. Responses to phytohemagglutinin (PHA) and alloantigen in a mixed lymphocyte culture (MLC) were repeatedly normal. It is interesting that the biotin concentration in standard lymphocyte culture media is equivalent to serum biotin levels achieved with therapy. It is possible that the normal in vitro lymphocyte findings were a result of this biotin-supplemented media. However, the ability of the patient's cells to respond to candida antigen in vitro was not corrected. This finding was consistent with the fact that in vivo delayed hypersensitivity skin tests to candida did not become positive for over a year following the initiation of biotin therapy.

The second child which we diagnosed as having MCD presented in the newborn period with ketoacidosis.[73] This child had several episodes of bacteremia associated with his illness. Lymphocytes were obtained immediately before biotin therapy was begun, and cultured with PHA in media which had a limited biotin content as well as biotin-supplemented media (Table 5). The lymphocyte response was absent in the limited media whereas it was restored to normal in the supplemented media. Since the child has been on pharmacologic doses of biotin we have been unable to repeat these studies. Subsequently, there has been a brief report of a child with PCC deficiency who had abnormal lymphocyte response to PHA when his serum was used as culture media which corrected when control serum was used.[56] This suggested that a toxic metabolite may be important in decreasing cellular immunity.

Table 5
LYMPHOCYTE RESPONSE TO
PHYTOHEMAGGLUTININ IN
NEONATAL MCD

	Culture medium[a]	
	+ Biotin	− Biotin
Culture day 5		
Patient	14008 ± 2677[b]	921 ± 347
Control	5663 ± 2559	5288 ± 1612
Culture day 6		
Patient	16145 ± 3048	386 ± 110
Control	—	9430 ± 1692

[a]   + Biotin = RPMI culture media; − Biotin = modified Eagle's medium.

[b] H^3-thymidine incorporation into DNA in PHA-stimulated peripheral blood lymphocytes in normal control and effected patient prior to biotin therapy. Results expressed as mean ± 1 S.D. cpm of 6 replicate cultures.

Other patients with MCD have had either candida dermatitis or bacterial or viral infections associated with their disease. There has been no specific pattern of immunodeficiency which has developed in MCD. Unfortunately, this is probably because most evaluations have been incomplete. For example, quantitative immunoglobulins have not been reported. Also, when in vitro lymphocyte tests have been performed it is likely that they were done in standard lymphocyte culture media which normally is supplemented with a large amount of biotin.

There is one other detailed immunologic study of a child with infantile onset MCD who had a clinical picture similar to our first reported case.[80,97] This patient's peripheral blood mononuclear cells were found to lack any in vitro suppressor activity. This included both T lymphocyte suppression of lymphocyte function as well as monocyte suppressor activity. Further evaluation revealed a defect in monocyte prostaglandin E_2 (PGE_2) production. Incubation of T cells with PGE_2 restored the T lymphocyte suppression. This child also had a low plasma level of linoleic acid, an essential fatty acid. A reasonable explanation for this latter finding is not available. Linoleic acid is converted to PGE_2 and it is possible that its conversion is dependent on malonyl-CoA, a product of ACC. What is more convincing of a primary relationship between MCD and these immune findings is that biotin therapy in this child corrected all of these defects.

There are several in vitro and in vivo animal models of biotin deficiency in which immune function has been evaluated. In the one in vitro system,[98] T lymphocyte cytotoxicity was evaluated using normal mouse spleen cells sensitized to P-815 target cells. The sensitization occurred over a 5-day culture period prior to the measurement of cytotoxicity. These investigators observed that when the culture medium used for the 5-day sensitization period consisted of dialyzed fetal calf serum (FCS) rather than undialyzed FCS, subsequent cytotoxicity was significantly depressed. Although biotin levels were not measured, biotin supplementation of the dialyzed FCS completely normalized the cytotoxicity, while avidin produced inhibitory effects similar to using dialyzed FCS. Finally, when a mixture of long chain fatty acids or palmitic acid alone (product of *de novo* fatty acid synthesis) was added the inhibitory effect of avidin was significantly reduced. The possibility of a toxic mechanism causing these results was not explored. These data are particularly interesting since they not only demonstrate a significant effect of biotin deficiency on one specific aspect of T cell immunity, but also indicate that this effect can occur quite rapidly, at least in a rodent system. We have been unable to demonstrate a similar effect of avidin and/or dialyzed human plasma on the human T cell response to PHA (unpublished data). It has been our experience and that of others that it is difficult to make normal human cells biotin deficient using short-term cultures of less than 10 days. If the findings in this mouse system can be extended to humans then at least two aspects of T cell immunity will have been demonstrated to be dependent on biotin, i.e., cytotoxicity and monocyte-mediated T cell suppression.

The possible role of fatty acid metabolism in T cell-mediated cytotoxicity is extremely interesting since it indicates one pathway which deserves further investigation. Other investigators have found little effect of unsaturated fatty acids on human natural cytotoxicity, an in vitro system which does not involve a sensitization step.[99] However, in one study[100] it was shown that ConA-stimulated human T cells require palmitic and oleic acid for optimal growth and in two other studies[101,102] *de novo* synthesis of fatty acids in human lymphocytes was significantly increased by lectin stimulation. Of particular interest, in thymocytes, 50% of the newly synthesized fatty acids came from the *de novo* process.[102] These results at least add some support for the mouse T cell cytotoxicity data.

Other studies of the effects of biotin deficiency on immunity have used various animal models. Early investigations found that biotin-deficient chicks and ducks had a more prolonged course and higher infection rate with three species of avian malaria, including *Plasmodium lophurae* than controls.[103,104] Even a mild degree of biotin deficiency increased susceptibility. Also, it was found that during infection of control birds, plasma and red cell biotin levels rose to a peak level coincident with the peak parasite level. Biotin supplementation of normal chicks had no antimalarial effect.[104] Finally, rats and mice which are biotin deficient have diminished resistance to oral infection with *Salmonella typhimurium*.[105] Spontaneous salmonella infection was also more frequent and virulent in the biotin-deficient animals compared to controls.

More recent studies in biotin-deficient rats have examined the ability to synthesize specific antibodies.[106-108] Biotin deficiency results in a significant depression of the antibody response to in vivo immunization with diphtheria toxin.[106] Thiamin deficiency in these experiments did not result in impaired antibody responses. Using a plaque-forming assay, these investigators found that the number of spleen antibody forming cells was significantly reduced in biotin-deficient rats immunized with sheep erythrocytes.[108] Furthermore, biotin therapy prior to immunization partially, but not completely restored antibody responses to normal. It is important to note that in these experiments, proper controls were used in which control animals were pair-fed the same diet as the experimental group but received biotin supplementation.

In summary, there is clinical and experimental evidence to indicate that biotin deficiency has significant effects on both T and B cell immunity. The precise functions which are affected are still being defined and any differences among species remain to be elucidated. The pathogenesis of immune deficiency in biotin-deficient states is not understood although evidence exists for both toxic as well as deficiency mechanisms. Finally, it is not known which biotin-dependent pathway, if any, is most important to maintain normal immunity. Recent reports implicate both PCC and ACC as being important to lymphocyte function. Interestingly, there are at least two reported patients with methylmalonic acidemia resulting from an inherited or acquired deficiency of B_{12}-dependent methylmalonyl-CoA mutase (the step following PCC) who had persistent candida infections.[109,110] Further studies are necessary to define the relative role of these various biotin-dependent pathways in lymphocyte metabolism.

VI. SUMMARY

Since biotin was first recognized as the ''egg white injury factor'' in 1936, interest in this B vitamin has changed from its function as a cofactor to its role in human diseases. In the past 20 years it has become clear that there are at least four mammalian metabolic pathways which are biotin dependent and which affect carbohydrate and amino and fatty acid metabolism. In addition, there is unproven speculation that biotin-dependent steps may be involved in *de novo* purine and pyrimidine synthesis and that somehow biotin is involved in histone, RNA, and protein metabolism.[111,112]

The details of the structure of biotin, its binding to apoenzyme, and the probable reactions involved in carboxylation are fairly well known. What is not known in sufficient detail is how or where biotin is absorbed in the GI tract; the significance, if any, of the bacterial synthesis of biotin; the mechanism of transport of biotin in the blood and across cell membranes; the location and steps in the binding of biotin to the apocarboxylases by its synthetase; the intracellular turnover, storage, and catabolism of biotin; and finally the variable dependency of different tissues on biotin-dependent steps. Until many of these processes are better understood it will be difficult to fully explain some of the manifestations of biotin deficiency.

Of particular interest and importance are the questions of how and why some organs and even areas within organs have different levels of carboxylases and presumably different dependencies on biotin. In fact, it appears that within a cell, some carboxylases are more affected by biotin deficiency than others. Is this due to different kinetic properties of the holosynthetase and/or different rates of turnover and/or other unknown factors? For example, is there a difference in biotin dependency during stages of cell maturation or differentiation? Are cells which are frequently dividing more biotin dependent than resting cells? Based on existing information one could offer possible answers to these questions but this would be only speculation as very little data exist.

Two manifestations of biotin-deficient and biotin-dependent states are severe and sometimes life-threatening: CNS disease and infection. If diagnosed and treated early many of these abnormalities are reversible. More follow-up is needed to completely determine long-term consequences. There is little basic understanding of the pathogenesis of the neuropathology in MCD. There are a paucity of studies using experimental models evaluating the effects of biotin deficiency on brain biochemistry. Those which have been reported have failed to establish reasonable explanations. As for the immune system, some information is accumulating as to the types of lymphocytes and functions which are affected in MCD. In addition, there is some information as to what the effect is of some of the different carboxylases. However, a great deal more work needs to be done to better define the relative importance of the different carboxylases and also the relative effect of MCD on different parts of the immune system, e.g., at different stages of maturation and differentiation. These studies are important not so much to help the children with MCD since most respond to biotin, but to characterize differences in metabolism between cell types which may ultimately allow us to selectively manipulate individual cell functions at the biochemical level.

Discovery of the inborn errors of biotin metabolism resulting in MCD is a reflection of significant technological advances which have been made in detecting and identifying various metabolites involved in carbohydate and amino and fatty acid metabolism. It is important that other defects in biotin metabolism be considered in children presenting with MCD. Also, these inborn errors provide a model for diagnosing and treating other vitamin-dependent disorders, both in the child and in the fetus. While a great deal has been accomplished since the discovery of the first patient in 1970, there are many questions which remain to be answered.

ACKNOWLEDGMENTS

I gratefully acknowledge Ms. Carol Dahlstrom for her outstanding typing skills and Seymour Packman, M.D. for his editorial and substantive comments. This work was in part supported by the Pediatric Clinical Research Center, in particular the Clinical Associate Physician Program (NIH #MO1 RR00079-18) and the March of Dimes Foundation (#6-328).

REFERENCES

1. **Moss, J. and Lane, M. D.**, The biotin-dependent enzymes, in *Advances in Enzymology*, Vol. 35, Meister, A., Ed., Interscience, New York, 1971, 321.
2. **Wood, H. G. and Barden, R. E.**, Biotin enzymes, *Ann. Rev. Biochem.*, 46, 385, 1977.
3. **Kogl, F. and Tonnis, B.**, in *Z. Physiol. Chem.*, 242, 43, 1936.
4. **Gardner, J., Parsons, H. T., and Peterson, W. H.**, Human biotin metabolism on various levels of biotin intake, *Arch. Biochem.*, 8, 339, 1945.
5. **Scott, D.**, Clinical biotin deficiency ("egg white injury"), *Acta Med. Scand.*, 162, 69, 1958.
6. **Baugh, C. M., Malone, J. H., and Butterworth, C. E., Jr.**, Human biotin deficiency: a case history of biotin deficiency induced by raw egg consumption in a cirrhotic patient, *Am. J. Clin. Nutr.*, 21(2), 173, 1968.
7. **Gompertz, D., Draffan, G. H., Watts, J. L., and Hull, D.**, Biotin-responsive β-methylcrotonylglycinuria, *Lancet*, July 3, 1971, 22.
8. **Sebrel!, W. H. and Harris, R. S., Eds.**, *The Vitamins*, Academic Press, New York, 1968, 261.
9. **MacLeod, P. R. and Lardy, H. A.**, Metabolic functions of biotin. II. The fixation of carbon dioxide by normal and biotin-deficient rats, *J. Biol. Chem.*, 179, 733, 1949.
10. **Wellner, V. P., Santos, J. I., and Meister, A.**, Carbamyl phosphate synthetase. A biotin enzyme, *Biochemistry*, 7, 2848, 1968.
11. **Feldott, G. and Lardy, H. A.**, Metabolic functions of biotin. IV. The role of carbamyl-L-glutamic acid in the synthesis of citrulline by normal and biotin-deficient rats, *J. Biol. Chem.*, 192, 447, 1951.
12. **Utter, M. F. and Keech, D. B.**, Formation of oxaloacetate from pyruvate and CO_2, *J. Biol. Chem.*, 235, PC17, 1960.
13. **Spencer, R. P. and Brody, K. R.**, Biotin transport by small intestine of rat, hamster, and other species, *Am. J. Physiol.*, 206, 653, 1964.
14. **Berger, E., Long, E., and Semenza, G.** The sodium activation of biotin absorption in hamster small intestine, *in vitro, Biochim. Biophys.*, 255, 873, 1972.
15. **Rose, R. C.**, Water-soluble vitamin absorption in intestine, *Ann. Rev. Physiol.*, 42, 157, 1980.
16. **Frank, O., Luisada-Opper, A. V., Feingold, S., and Baker, H.**, Vitamin binding by human and some animal plasma proteins, *Nutr. Rep. Int.*, 1(3), 161, 1970.
17. **Horsburgh, T. and Gompertz, D.**, A protein-binding assay for measurement of biotin in physiological fluids, *Clin. Chem. Acta*, 82, 215, 1978.
18. **Oppel, T. W.**, Studies of biotin metabolism in man. I. The excretion of biotin in human urine, *Am. J. Med. Sci.*, 204, 856, 1942.
19. **Gardner, J., Parsons, H. T., and Peterson, W. H.**, Human utilization of biotin from various diets, *Am. J. Med. Sci.*, 211, 198, 1946.
20. **Denko, C., Grundy, W., Porter, I., Berryman, G., Friedmann, T., and Youmans, J.**, Excretion of B complex vitamins in urine and feces of 7 normal adults, *Arch. Biochem.*, 10, 33, 1946.
21. **Lee, H.-M., Wright, L. D., and McCormick, D. B.**, Metabolism of carbonyl-labeled ^{14}C-biotin in the rat, *J. Nutr.*, 102, 1453, 1972.
22. **Lee, H.-M., McCall, N. E., Wright, L. E., and McCormick, D. B.**, Urinary excretion of biotin and metabolites in the rat, *Proc. Soc. Exp. Biol. Med.*, 142, 642, 1973.
23. **Dakshinamurti, K. and Mistry, S. P.**, Tissue and intracellular distribution of biotin-C^{14} OOH in rats and chicks, *J. Biol. Chem.*, 238, 294, 1963.
24. **Siegel, S., Nixon, W. E., and Dhyse, F. D.**, Distribution of biotin and tritiated biotin in maternal, prenatal and postnatal dog tissues, *Growth*, 13, 61, 1967.
25. **Anderson, J. K. S.**, Characterization of the biotin transport system and study of the requirement for biotin in LM cells, a mammalian cell line, *Diss. Abst. Int. B:*, 35, 2325, 1974.
26. **Petrelli, F., Moretti, P., and Paparelli, M.** Intracellular distribution of biotin ^{14}C OOH in rat liver, *Molec. Biol. Rep.*, 4(4), 247, 1978.
27. **Koscow, D. P., Huang, S. C., and Lane, M. D.**, Propionyl holocarboxylase synthesis I. Preparation and properties of the enzyme system, *J. Biol. Chem.*, 237, 3633, 1962.

28. **Achuta Murthy, P. N. and Mistry, S. P.,** *In vitro* synthesis of propionyl Co-A holocarboxylase by a partially purified mitochondrial preparation from biotin-deficient chicken liver, *Can. J. Biochem.,* 52, 800, 1974.

29. **Burri, B. J., Sweetman, L., and Nyhan, W. L.,** Mutant holocarboxylase synthetase, *J. Clin. Invest.,* 68, 1491, 1981.

30. **Kluger, R. and Adaqadkar, P. D.,** A reaction proceeding through intramolecular phosphorylation of a urea. A chemical mechanism for enzymatic carboxylation of biotin involving cleavage of adenosine 5′-triphosphate, *J. Am. Chem. Soc.,* 98, 3741, 1976.

31. **Ashman, L. K. and Keech, D. B.,** Sheep kidney pyruvate carboxylase, *J. Biol. Chem.,* 250, 14, 1975.

32. **Iwahara, S., McCormick, D., Wright, G., and Li, H.,** Bacterial degradation of biotin, *J. Biol. Chem.,* 244, 1393, 1969.

33. **Sydenstricker, V. P., Singal, S. A., Briggs, A. P., DeVaugh, N. M., and Isbell, H.,** Observations on the "egg white injury" in man, *JAMA,* 118, 1199, 1942.

34. **Williams, R. H.,** Clinical biotin deficiency, *N. Engl. J. Med.,* 228; 274, 1943.

35. **Sweetman, L., Surh, L., Baker, H., Peterson, R. M., and Nyhan, W. L.,** Clinical and metabolic abnormalities in a boy with dietary deficiency of biotin, *Pediatrics,* 68, 55, 1981.

36. **Mock, D. M., DeLorimier, A. A., Liebman, W. M., Sweetman, L., and Baker, H.,** Biotin deficiency: an unusual complication of parenteral alimentation, *N. Engl. J. Med.,* 304, 820, 1981.

37. **Kien, C. L., Kohler, E., Goodman, S, I., Berlow, S., Hong, R., Horowitz, S., and Baker, H.,** Biotin-responsive *in vivo* carboxylase deficiency in two siblings with secretory diarrhea receiving total parenteral nutrition, *J. Pediatr.,* 99, 546, 1981.

38. **Nisenson, A.,** Seborrheic dermatitis of infants and Leiner's disease: a biotin deficiency, *J. Pediatr.,* 5, 537, 1957.

39. **Miller, M. E. and Koblenzer, P. J.,** Leiner's disease and deficiency of C5, *J. Pediatr.,* 80, 879, 1972.

40. **Svehcar, J. and Homolka, J.,** Experimental experiences with biotin in babies, *Ann. Paediatr.,* 174, 175, 1950.

41. **Barker, L. P., Gross, P., and McCarthy, J. T.,** Erthyrodermas of infancy, *Arch. Dermatol.,* 77, 201, 1958.

42. **Keipert, J. A.,** Oral use of biotin in seborrhoeic dermatitits of infancy: a controlled trial, *Med. J. Aust.,* 1, 584, 1976.

43. **Erlichman, M., Goldstein, R., Levi, E., Greenberg, A., and Freier, S.,** Infantile flexural seborrheic dermatitis. Neither biotin nor essential fatty acid deficiency, *Arch. Dis. Child.,* 56, 560, 1981.

44. **Goldsmith, L. A.,** Vitamins and alopecia, *Arch. Dermatol.,* 116, 1135, 1980.

45. **Barlow, G. B., Dickerson, J. A., and Wilkinson, A. W.,** Plasma biotin levels in children with burns and scalds, *J. Clin. Pathol.,* 29, 58, 1976.

46. **Johnson, A. R. and Hood, R. L.,** Biotin and the sudden infant death syndrome, *Nature (London),* 285, 159, 1980.

47. **Hommes, F. A., Kuipers, J. R. G., Elema, J. D., Jansen, J. F., and Jonxis, J. H. P.,** Propionicaciduria, a new inborn error of metabolism, *Pediatr. Res.,* 2, 519, 1968.

48. **Blom, W., de Muinckkeizer, S., and Scholte, H. R.,** Acetyl CoA carboxylase deficiency. An inborn error of *de novo* fatty acid synthesis, *N. Engl. J. Med.,* 305(8), 465, 1981.

49. **Weyler, W., Sweetman, L., Maggio, D. C., and Nyhan, W. L.,** Deficiency of propionyl-CoA carboxylase and methylcrotonyl-CoA carboxylase in a patient with methylcrotonylglycinuria, *Clin. Chim. Acta,* 76, 321, 1977.

50. **Sweetman, L., Bates, S. P., Hull, D., and Nyhan, W. L.,** Propionyl-CoA carboxylase deficiency in a patient with biotin-responsive 3-methylcrotonylglycinuria, *Pediatr. Res.,* 11, 1144, 1977.

51. **Roth, K., Coh, R., Yandrasitz, B. S., Preti, G., Dodd, P., and Segal, S.,** Beta-methylcrotonic aciduria associated with lactic acidosis, *J. Pediatr.,* 88, 229, 1976.

52. **Cowan, M. J., Packman, S., Wara, D. W., Ammann, A. J., Yoshino, M., Sweetman, L., and Nyhan, W.,** Multiple biotin-dependent carboxylase deficiencies associated with defects in T-cell and B-cell immunity, *Lancet,* July 21, 1979, 115.

53. **Wolf, B., Hsia, Y. E., Sweetman, L., Gravel, R., Harris, D. J., and Nyhan, W. L.,** Propionic acidemia: a clinical update, *J. Pediatr.,* 99(6), 835, 1981.

54. **Brandt, I. K., Hsia, Y. E., Clement, D. H., and Provence, S. A.,** Propionic-acidemia dietary treatment resulting in normal growth and development, *Pediatrics,* 53, 391, 1974.

55. **Wolf, B., Paulsen, E. P., and Hsia, Y. E.,** Asymptomatic propionyl CoA carboxylase deficiency in a 13-year-old girl, *J. Pediatr.,* 95, 563, 1979.

56. **Muller, S., Falkenberg, N., Monch, E., and Jacobs, C.,** Propionacidaemia and immunodeficiency, *Lancet,* May 8, 1980, 551.

57. **Wadlington, W. B., Kilroy, A., Ando, T., Sweetman, L., and Nyhan, W. L.,** Hyperglycinemia and propionyl CoA carboxylase deficiency and episodic severe illness without consistent ketosis, *J. Pediatr.,* 86, 707, 1975.

58. Naylor, G., Sweetman, L., Nyhan, W. L., Hornbeck, C., Griffiths, J., Morch, L., and Brandage, S., Isotope dilution analysis of methylcitric acid in amniotic fluid for the prenatal diagnosis of propionic and methylmalonic acidemia, *Clin. Chim. Acta*, 107, 175, 1980.

59. Wolf, B., Willard, H. F., and Rosenberg, L. E., Kinetic analysis of genetic complementation in heterokaryons of propionyl CoA carboxylase-deficient human fibroblasts, *Am. J. Hum. Genet.*, 32, 16, 1980.

60. Wolf, B. Molecular basis for genetic complementation in propionyl CoA carboxylase deficiency, *Exp. Cell Res.*, 125, 502, 1980.

61. Kidd, J. R., Wolf, B., Hsia, Y. E., and Kidd, K. K., Genetics of propionic acidemia in a Mennonite-Amish kindred, *Am. J. Hum. Genet.*, 32, 236, 1980.

62. Wolf, B. Reassessment of biotin-responsiveness in "unresponsive" propionyl CoA carboxylase deficiency, *J. Pediatr.*, 97, 964, 1980.

63. Stokke, O., Eldjarn, L., Jellum, E., Pande, H., and Waaler, P. E., Beta-methylcrotonyl-CoA carboxylase deficiency: a new metabolic error in leucine degradation, *Pediatrics*, 59, 726, 1972.

64. Gompertz, D., Bartlett, K., Blair, D., and Stern, C., Child with a defect in leucine metabolism associated with β-hydroxyisovaleric aciduria and β-methylcrotonylglycinuria, *Arch. Dis. Child.*, 48, 975, 1973.

65. Landaas, S., Accumulation of 3-hydroxyisobutyric acid, 2-methyl-3-hydroxy-butyric acid and 3-hydroxyisovaleric acid in ketoacidosis, *Clin. Chim. Acta*, 64, 143, 1975.

66. Hommes, F. A., Polman, H. A., and Reerink, J. D., Leigh's encephalopathy: an inborn error of gluconeogenesis, *Arch. Dis. Child.*, 43, 423, 1968.

67. Tada, K., Yoshida, T., Kouno, T., Wada, Y., Yokoyama, A., and Arakawa, T., Hyperalaninemia and pyruvicemia, *Tohoku J. Exp. Med.*, 22, 99, 1969.

68. Vidailhet, M., Lefebvre, E., Beley, G., and Marsa, C., Neonatal lactic acidosis with pyruvate carboxylase inactivity, *J. Inherited Metab. Dis.*, 4, 131, 1981.

69. Atkin, B. M., Buist, N. R. M., Utter, M. F., Leiter, A. B., and Banker, B. Q., Pyruvate carboxylase deficiency and lactic acidosis in a retarded child without Leigh's disease, *Pediatr. Res.*, 13, 109, 1979.

70. Murphy, J. V., Isohashi, G., Weinberg, M. B., and Utter, M. F., Pyruvate carboxylase deficiency: an alleged biochemical cause of Leigh's disease, *Pediatrics*, 68, 401, 1981.

71. Atkin, B. M., Carrier detection of pyruvate carboxylase deficiency in fibroblasts and lymphocytes, *Pediatr. Res.*, 13, 1101, 1979.

72. Wolf, B., Hsia, E., Sweetman, L., Feldman, G., Boychuk, R. B., Bart, R. D., Crowell, D. H., Di Mauro, R. M., and Nyhan, W. L., Multiple carboxylase deficiency: clinical and biochemical improvement following neonatal biotin treatment, *Pediatrics*, 68, 113, 1981.

73. Packman, S., Sweetman, L., Baker, H., and Wall, S., The neonatal form of biotin-responsive multiple carboxylase deficiency, *J. Pediatr.*, 99, 418, 1981.

74. Roth, K. S., Cohn, R., Yandrasitz, J., Preti, G., Dodd, P., and Segal, S., Beta-methylcrotonyl aciduria associated with lactic acidosis, *J. Pediatr.*, 88, 229, 1976.

75. Roth, K. S., Yand, W., Allan, L., Saunders, M., Gravel, R. A., and Dakshinamurti, K., Prenatal administration of biotin in biotin responsive multiple carboxylase deficiency, *Pediatr. Res.*, 16, 126, 1982.

76. Saunders, M. E., Sherwood, W. G., Duthie, M., Suhr, L., and Gravel, R. A., Evidence for a defect of holocarboxylase synthetase activity in cultured lymphoblasts from a patient with biotin-responsive multiple carboxylase deficiency, *Am. J. Hum. Genet.*, in press.

77. Bartlett, K., Ng, H., and Leonard, J. V., A combined defect of three mitochondrial carboxylases presenting as biotin-responsive 3-methylcrotonyl glycinuria and 3-hydroxyisovaleric aciduria, *Clin. Chim. Acta*, 100, 183, 1980.

78. Thoene, J., Baker, H., Yoshino, M., and Sweetman, L., Biotin-responsive carboxylase deficiency associated with subnormal plasma and urinary biotin, *N. Engl. J. Med.*, 304, 817, 1981.

79. Charles, B. M., Hosking, G., Green, A., Pollitt, R., Bartlett, K., and Taitz, L., Biotin-responsive alopecia and developmental regression, *Lancet*, July 21, 1979, 118.

80. Munnich, A., Saudubray, J. M., Coudé, F. X., Ogier, H., Charpentier, C., Marsae, C., Carre, G., Bourgeay-Causse, M., and Frezal, J., Biotin dependent multiple carboxylase deficiency presenting as a congenital lactic acidosis, *Eur. J. Pediatr.*, 137, 203, 1981.

81. Packman, S., Caswell, N. M., and Baker, H., Biochemical evidence for diverse etiologies in biotin-responsive multiple carboxylase deficiency, *Biochem. Genet.*, 20, 17, 1982.

82. Packman, S., Sweetman, L., Yoshino, M., Baker, H., and Cowan, M., Biotin-responsive multiple carboxylase deficiency of infantile onset, *J. Pediatr.*, 99, 421, 1981.

83. Sweetman, L., Two forms of biotin-responsive multiple carboxylase deficiency, *J. Inherited Metab. Dis.*, 4, 1981.

84. O'Connor, L. R., Klein, K. L., and Bethune, J. E., Hyperphosphatemia in lactic acidosis, *N. Engl. J. Med.*, 297, 707, 1977.

85. Tada, K., Corbeel, L. M., Eeckels, R., and Eggermont, E., A block in glycine cleavage reaction as a common mechanism in ketotic and nonketotic hyperglycinemia, *Pediatr. Res.*, 8, 721, 1974.

86. **Packman, S., Cowan, M. J., Golbus, M. S., Caswell, N. M., Sweetman, L., Burri, B. J., Nyhan, W. L., and Baker, H.,** Prenatal treatment of biotin-responsive multiple carboxylase deficiency, *Lancet,* June 26, 1982, 1435.

87. **Bartlett, K., and Gompertz, D.,** Combined carboxylase defect: biotin-responsiveness in cultured fibroblasts, *Lancet,* October 9, 1976, 804.

88. **Saunders, M., Sweetman, L., Robinson, B., Roth, K., Cohn, R., and Gravel, A.,** Biotin-response organicaciduria, *J. Clin. Invest.,* 64, 1695, 1979.

89. **Munnich, A., Saudubray, J. M., Carre, G., Coudé, F. X., Ogier, H., Charpentier, C., and Frezal, J.,** Defective biotin absorption in multiple carboxylase deficiency, *Lancet,* August 1, 1981, 263.

90. **Feldman, G. L. and Wolf, B.,** Deficient acetyl CoA carboxylase activity in multiple carboxylase deficiency, *Clin. Chim. Acta,*111, 147, 1981.

91. **Adibi, S. A.,** Roles of branched-chain amino acids in metabolic regulation, (editorial review), *J. Lab. Clin. Med.,* 95, 475, 1980.

92. **Sander, J. E., Malamud, N., Cowan, M. J., Packman, S., Ammann, A. J., and Wara, D. W.,** Intermittent ataxia and immunodeficiency with multiple carboxylase deficiencies: a biotin-responsive disorder, *Ann. Neurol.,* 8, 544, 1980.

93. **Rosenberg, R. N.,** Biochemical genetics of neurologic disease, *N. Engl. J. Med.,* 305, 1181, 1981.

94. **Israels, S., Haworth, J. C., Dunn, H. G., and Applegarth, D. A.,** Lactic acidosis in childhood, *Adv. Pediatr.,* 22, 267, 1976.

95. **Schrijver, J., Dias, T., and Hommes, F. A.,** Some biochemical observations on biotin deficiency in the rat as a model for human pyruvate carboxylase deficiency, *Nutr. Metab.,* 23, 179, 1979.

96. **Sander, J. E., Packman, S., and Townsend, J. J.,** Brain pyruvate carboxylase and the pathophysiology of biotin-dependent diseases, *Neurology,* 32, 878, 1982.

97. **Munnich, A., Fischer, A., Saudubray, J. M., Griscelli, C., Coudé, F. X., Ogier, H., Charpentier, C., and Frezal, J,.** Biotin-responsive immunoregulatory dysfunction in multiple carboxylase deficiency, *J. Inherited Metab. Dis.,* 4, 113, 1981.

98. **Kung, J. T., Mackenzie, C. G., and Talmage, D. W.,** The requirement for biotin and fatty acids in the cytotoxic T-cell response, *Cell. Immunol.,* 48, 100, 1979.

99. **Rice, C., Hudig, D., Newton, R. S., and Mendelsohn, J.,** Effect of unsaturated fatty acids on human lymphocytes, *Clin. Immunol. Immunopathol.,* 20, 389, 1981.

100. **Spieker-Polet, H. and Polet, H.,** Requirement of a combination of a saturated and an unsaturated free fatty acid and a fatty acid carrier protein for *in vitro* growth of lymphocytes, *J. Immunol.,* 126, 949, 1981.

101. **Blomstrand, R.,** Fatty acid synthesis in human lymphocytes, *Acta Chem. Scand.,* 20, 1122, 1966.

102. **Liljeqvist, L.,** Lipid biosynthesis in human thoracic duct lymphocytes and thymocytes, *Acta Chem. Scand.,* 27, 891, 1973.

103. **Seeler, A. O., Ott, W. H., and Gundel, M. E.,** Effect of biotin deficiency on the course of *Plasmodium lophurae* infection in chicks, *Proc. Soc. Biol. Med.,* 55, 107, 1944.

104. **Trager, W.,** The influence of biotin upon susceptibility to malaria, *J. Exp. Med.,* 77, 557, 1943.

105. **Kliger, I. J.,** Nutritional deficiency and resistance to infection; effect of biotin deficiency on susceptibility of rats and mice to infection with *Salmonella typhimurium, J. Infect. Dis.,* 78, 60, 1946.

106. **Pruzansky, J. and Axelrod, A. E.,** Antibody production to diphtheria toxoid in vitamin deficiency states, *Soc. Exp. Biol. Med.,* 89, 323, 1955.

107. **Axelrod, A. E.,** Immune processes in vitamin deficiency states, *Am. J. Clin. Nutr.,* 24, 265, 1971.

108. **Kumar, M. and Axelrod, A. E.,** Cellular antibody synthesis in thiamin, riboflavin, biotin and folic acid-deficient rats, *Soc. Exp. Biol. Med.,* 157, 421, 1978.

109. **Higginbottom, M. D., Sweetman, L., and Nyhan, W. L.,** A syndrome of methylmalonic aciduria, homocystinuria, megaloblastic anemia and neurologic abnormalities in a vitamin B_{12}-deficient breast-fed infant of a strict vegetarian, *N. Engl. J. Med.,* 229, 317, 1978.

110. **Nyhan, W. L., Fawcett, N., and Anindo, T.,** Response to dietary therapy in B_{12} unresponsive methylmalonic acidemia, *Pediatrics,* 51, 539, 1973.

111. **Boeckx, R. L. and Dakshinamurti, K.,** Effect of biotin on ribonucleic acid synthesis, *Biochim. Biophys. Acta,* 383, 282, 1975.

112. **Petrelli, F., Coderoni, S., Moretti, P., and Paparelli, M.,** Effect of biotin on phosphorylation, acetylation, methylation of rat liver histones, *Molec. Biol. Rep.,* 4, 87, 1977.

Chapter 2

SYNERGISTIC STIMULATORY EFFECTS OF VITAMIN D_3 METABOLITES AND PARATHYROID HORMONE (PTH) UPON BONE CALCIFICATION IN VITRO

Mamoru Kiyoki, Kohtaro Kawashima, and Hiroyoshi Endo

TABLE OF CONTENTS

I. INTRODUCTION

Homeostasis of the structure and metabolism of living bones depends upon a fine balance between bone formation by osteoblasts and bone resorption by osteoclasts. Several physiological factors involved in the processes of bone resorption have been clearly demonstrated by measuring the amount of ^{45}Ca released in vitro from prelabeled fetal rat bones having high activities of resorption as well as formation.[1-3] However, little is known of the physiological factors responsible for bone formation, because no bone culture system has been successfully applied for clarifying the factors under defined in vitro conditions. For example, while vitamin D has long been known as an anti-rachitic agent for stimulating bone formation in living animals, it is not yet clear whether the effect is due directly to enhanced calcium deposition of the bone or indirectly due to increased calcium absorption from the intestine.[4] The problem could be resolved by another type of bone culture system in which formation of the organic bone matrix and its calcification preferentially proceeds in vitro without dissolution of the newly formed bone.

A number of studies using organ culture of bones has been carried out since Fell and Robison (1929)[5] reported that normal calcified bone was formed after 1 month of culturing 5-day chick embryonic cartilaginous rudiments of the femora on a solid medium. In these studies most workers have used media containing natural biological fluids such as plasma and serum so they could describe the process of bone formation, but not analyze the factors involved in it. Along these lines we set up organ cultures in liquid media of 9-day chick embryonic femora which displayed the initial signs of periosteal calcification and found that physiological diaphyseal calcification can be obtained after 6 days of cultivation only when the concentration of serum in the medium was increased up to 50% in the presence of 10% 11-day chick embryo extract.[6,7] These findings indicated that serum contains physiological factors which are prerequisites for the indication of bone calcification.

In living bone, on the one hand, osteoblastic and osteoclastic cells are in close proximity to each other and are actively engaged in opposite functions of bone formation and resorption, all in the same humoral environment. From such a homeostatic situation in the bone it would be reasonable to postulate that osteoblasts and osteoclasts both have their activities regulated by the same physiological factors in the body fluid. On this point, PTH and 1α, 25-dihydroxy vitamin D_3 (1α,25-$(OH)_2D_3$) have been repeatedly confirmed to stimulate the process of bone resorption.[1-3] Moreover, 24R,25-dihydroxy vitamin D_3 (24R, 25-$(OH)_2D_3$) recently has been suggested to be involved in bone resorption induced by PTH[8] and to stimulate bone formation as well.[9-11] Thus we supposed that PTH and vitamin D_3 metabolites are possible candidates for bone formation-stimulating factors in living animals.

On the other hand, bone formation consists of a long series of processes: (1) proliferation of osteoblasts; (2) elaboration by these cells of intercellular organic matrix principally composed of collagen and chondroitin sulfate; and (3) deposition over the matrix of bone salt mostly composed of calcium phosphate. A different controlling factor or a different combination of controlling factors might be involved in each process, since these processes probably differ in mechanism. If such is the case, all the factors would have to be simultaneously added into the culture medium in order to get physiological calcification of the bone in vitro.

In this work, therefore, we have employed a bone culture system in which serum was replaced by an enriched, chemically defined medium and have examined vitamin D_3 metabolites and PTH added singly or in combination for their bone calcification-stimulating activity.

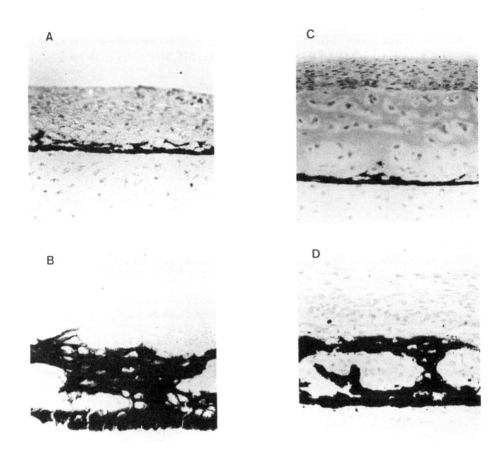

FIGURE 1. Diaphyses of the femora from 9- or 11-day chick embryos. (A) 9-day chick embryo femur; (B) 11-day chick embryo femur; (C) 9-day chick embryo femur cultured in basal medium for 6 days; (D) 9-day chick embryo femur cultured in basal medium containing $1\alpha,25\text{-}(OH)_2D_3$, $24R,25\text{-}(OH)_2D_3$, and PTH for 6 days.

II. MATERIALS AND METHODS

Starting materials for bone cultures were the femora from 9-day chick embryos. The bone is a cartilaginous rudiment covered with a thin layer of perichondrium. In the diaphyseal portion of the bone, however, an osteogenic layer with a small number of osteoblasts can be seen as a narrow calcified zone between the fibroblastic layer as the outermost and the cartilage mass as the central part of the bone (Figure 1A). In the periosteum of the developing bone, no sign of osteoclastic processes can be observed.

We primarily used the same culture system described previously.[6,7] Each femur was transferred to a roller-tube contaning 1.5 mℓ of medium. To allow for differences between embryos, the bones from one side of an embryo were cultivated in control medium, and those from the opposite side of the same embryos, in experimental medium. The basal medium was a mixture (9:1) of BGJb-HW2[12] and 11-day chick embryo extract. Roller-tubes were incubated at 37°C for 6 days and rotated 10 times per hour in a roller drum which held them at an angle of 5°. The medium was replaced every other day.

At the end of the cultivation period, the calcium of most pair-mated bones was chemically determined; the ratio of calcium levels in treated femora to their pair-mated control bones was calculated as the calcification ratio. Thus values of more than 1.00 show a stimulation of bone calcification by the treatment. Moreover, the histological development of residual

Table 1
CALCIUM LEVELS IN 9-DAY-OLD CHICK EMBRYONIC FEMORA AFTER 6 DAYS OF CULTIVATION USING BASAL MEDIUM AS CONTROL

| Treatment | Concentration | Calcium level (μg/femur) | | Calcification ratio[a] | No. of bone pairs |
		Control bones	Treated bones		
PTH	1 (unit/mℓ)	6.35 ± 1.00	6.89 ± 0.81	1.04 ± 0.13	11
$1\alpha,25$-$(OH)_2D_3$	0.02 (ng/mℓ)				
$+24R,25$-$(OH)_2D_3$	0.5	9.60 ± 1.15	10.27 ± 0.58	1.17 ± 0.15	7
$+25$-OH-D_3	2.5				
$1\alpha,25$-$(OH)_2D_3$	0.02				
$+24R,25$-$(OH)_2D_3$	0.5	6.45 ± 1.47	7.31 ± 1.43	1.14 ± 0.06	6
$1\alpha,25$-$(OH)_2D_3$	0.02	10.06 ± 0.30	10.12 ± 0.85	1.01 ± 0.07	5
$1\alpha,25$-$(OH)_2D_3$	2.0	9.84 ± 1.46	8.84 ± 1.46	0.93 ± 0.11	5
$24R,25$-$(OH)_2D_3$	0.5	10.40 ± 0.39	9.52 ±0.76	0.91 ± 0.05	5
$24R,25$-$(OH)_2D_3$	50.0	9.92 ± 0.87	8.76 ± 0.27	0.92 ± 0.10	5

Note: The values represent means ± standard error. The amount of calcium deposited in the femur was determined by the OPCP method.[20] Control femora were cultivated in basal medium.

[a]   The ratio of treated femora to pair-mated control bones in terms of calcium level.

pair-mated bones was microscopically examined with specimens stained with hematoxylin-eosin and von Kossa's stain.

III. RESULTS

A. A Combination of Physiological Factors, PTH and Vitamin D_3 Metabolites, as a Prerequisite for Inducing Bone Calcification In Vitro

In the first series of experiments (Table 1), the control medium was basal medium alone and the experimental medium was basal medium plus PTH or several analogues of vitamin D_3.

In medium containing PTH alone, calcification of the bones scarcely advanced either chemically or histologically during 6 days of cultivation, although the bones did elongate significantly.[14] Similarly, $1\alpha,25$-$(OH)_2D_3$ and $24R,25$-$(OH)_2D_3$ gave no stimulated calcification when administered singly or in combination with 25-OH-D_3. Femora cultured in these media formed only osteoid tissue as observed in control bone cultures (Figure 1C).

In the second series of experiments (Table 2), the control medium was basal medium with PTH added and the experimental medium was the control medium plus several analogues of vitamin D_3 or their combinations.

Combined treatment of the bones with $1\alpha,25$-$(OH)_2D_3$, $24R,25$-$(OH)_2D_3$, and 25-OH-D_3 in the presence of PTH gave a calcification ratio of 2.07, thus indicating highly stimulated bone mineralization. Without 25-OH-D_3, moreover, the other two kinds of dihydroxy vitamin D_3 metabolites in the presence of PTH caused much the same degree of stimulation of the calcification process. The histological features of the diaphyseal cortex of the treated bones were histologically confirmed to be similar to these of 11-day embryonic femora in the normal course toward hatching (Figure 1, B and D). When combined with 25-OH-D_3 in the presence of PTH, however, neither $1\alpha,25$-$(OH)_2D_3$ nor $24R,25$-$(OH)_2D_3$ enhanced calcification, the calcification ratios being 1.05 and 1.15, respectively. Moreover, no single treatment with $1\alpha,25$-$(OH)_2D_3$ or $24R,25$-$(OH)_2D_3$ in the presence of PTH affected calcification, even at high concentrations.

Table 2

**CALCIUM LEVELS IN 9-DAY-OLD CHICK EMBRYONIC FEMORA AFTER 6
DAYS OF CULTIVATION USING PTH-CONTAINING BASAL MEDIUM AS
CONTROL**

Treatment	Concentration (ng/mℓ)	Calcium level (μg/femur)		Calcification ratio[a]	No. of bone pairs
		Control bones	Treated bones		
1α,25-(OH)₂D₃ +24R,25-(OH)₂D₃ +25-OH-D₃	0.02 0.5 2.5	4.42 ± 0.74	7.54 ± 0.79	2.07 ± 0.16[b]	11
1α,25-(OH)₂D₃ +24R,25-(OH)₂D₃	0.02 0.5	4.40 ± 0.63	7.71 ± 0.60	2.04 ± 0.24[b]	13
1α,25-(OH)₂D₃ +25-OH-D₃	0.02 2.5	5.61 ± 0.56	5.71 ± 0.57	1.05 ± 0.09	10
24R,25-(OH)₂D₃ +25-OH-D₃	0.5 2.5	7.65 ± 0.69	8.93 ± 0.43	1.15 ± 0.11	8
1α,25-(OH)₂D₃	0.02	5.03 ± 0.79	5.55 ± 0.99	1.05 ± 0.06	6
1α,25-(OH)₂D₃	2.0	6.60 ± 1.12	6.96 ± 0.79	1.38 ± 0.25	5
24R,25-(OH)₂D₃	0.5	10.88 ± 1.21	11.13 ± 1.12	1.06 ± 0.05	5
24R,25-(OH)₂D₃	50.0	8.74 ± 1.21	8.14 ± 0.83	0.97 ± 0.08	5
25-OH-D₃	2.5	5.65 ± 1.74	5.80 ± 1.46	1.00 ± 0.13	5

Note: The values represent means ± standard error. Control bones were cultivated in basal medium with added PTH.

[a] The ratio of treated femora to pair-mated control bones in terms of calcium level.
[b] Statistical significance $p < 0.005$.

Our results above clearly demonstrate that vitamin D_3 metabolites, especially the combination of $1\alpha,25\text{-}(OH)_2D_3$ and $24R,25\text{-}(OH)_2D_3$, stimulate bone calcification under the synergistic influence of PTH.

B. Dose-Response Relationship of Calcification-Stimulating Factors

In the next series of experiments, the dose-response relationship of three calcification-stimulating factors was studied using the same bone culture system (Figure 2).

For examining dose-dependent effects of $1\alpha,25\text{-}(OH)_2D_3$, control bones were cultivated in PTH-containing basal medium; experimental bones were cultivated in the same medium to which varying amounts of $1\alpha,25\text{-}(OH)_2D_3$ were added in the presence or absence of $24R,25\text{-}(OH)_2D_3$. In the absence of $24R,25\text{-}(OH)_2D_3$, $1\alpha,25\text{-}(OH)_2D_3$ did not stimulate calcification at concentrations ranging from 0.02 to 2.0 ng/mℓ. In the presence of $24R,25\text{-}(OH)_2D_3$ (0.5 ng/mℓ), however, $1\alpha,25\text{-}(OH)_2D_3$ significantly accelerated bone mineralization only at a concentration of 0.02 ng/mℓ which corresponds to its physiological level (Figure 2A).

As for the dose-response relationship of $24R,25\text{-}(OH)_2D_3$, control bones were grown in PTH-containing basal medium and pair-mated treated bones in the same medium to which varying amounts of $24R,25\text{-}(OH)_2D_3$ were added in the presence or absence of $1\alpha,25\text{-}(OH)_2D_3$. In the absence of $1\alpha,25\text{-}(OH)_2D_3$, $24R,25\text{-}(OH)_2D_3$ gave no stimulation of bone mineralization. In the presence of $1\alpha,25\text{-}(OH)_2D_3$ (optimal at 0.02 ng/mℓ), however, $24R,25\text{-}(OH)_2D_3$ promoted calcification at all concentrations higher than 0.5 ng/mℓ which corre-

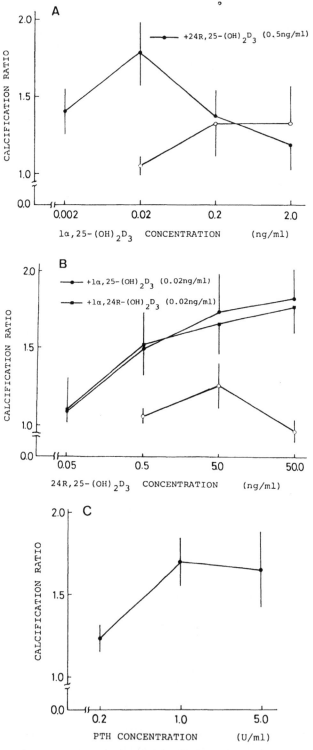

FIGURE 2.   Dose-response relationship of mineralization factors on calcium level of 9-day chick embryonic femur after 6 days of cultivation. (A) 1α,25-(OH)₂D₃; (B) 24R,25-(OH)₂D₃; (C) PTH. The control bones were cultivated in basal medium with added PTH (A and B) or 1α,25-(OH)₂D₃ and 24R,25-(OH)₂D₃ (C).

sponded to its physiological level. Furthermore, the stimulation became more striking with increasing concentration (Figure 2B).

As for the effect of PTH, varying amounts of the hormone were added to control medium in which concentrations of $1\alpha,25\text{-}(OH)_2D_3$ and $24R,25\text{-}(OH)_2D_3$ were established at 0.02 and 0.5 ng/mℓ, respectively. Within a range from 0.2 to 5.0 units/mℓ, 1.0 and 5.0 units/mℓ of PTH gave calcification ratios of 1.70 and 1.65, respectively, indicating a significant increase in calcium deposition in the bone (Figure 2C).

C. Effect on Calcification of Natural and Synthetic Vitamin D_3 Analogues Other Than $1\alpha,25\text{-}(OH)_2D_3$ and $24R,25\text{-}(OH)_2D_3$

Using the same bone culture system we examined the effects on calcification of a variety of natural metabolites and synthetic analogues of vitamin D_3. Of these substances tested we report here the effect of $1\alpha,24R,25\text{-}(OH)_3D_3$ and $1\alpha,24R\text{-}(OH)_2D_3$. The former is a metabolite which can be formed from both $1\alpha,25\text{-}(OH)_2D_3$ and $24R,25\text{-}(OH)_2D_3$; the latter is a synthetic analogue which has been confirmed to stimulate intestinal calcium absorption as does $1\alpha,25\text{-}(OH)_2D_3$.[15]

As to the effect of $1\alpha,24R,25\text{-}(OH)_3D_3$, this vitamin D_3 metabolite gave no stimulation of calcification when administered singly or in combination with $1\alpha,25\text{-}(OH)_2D_3$ in the presence of PTH. Only when combined with $24R,25\text{-}(OH)_2D_3$ in the presence of PTH, however, the metabolite did cause slight but significant enhancement of bone mineralization, suggesting that the trihydroxy metabolite is much lower, but similar, in activity as compared with $1\alpha,25\text{-}(OH)_2D_3$.

The synthetic vitamin D_3 analogue, $1\alpha,24R\text{-}(OH)_2D_3$, on the other hand, could completely replace $1\alpha,25\text{-}(OH)_2D_3$ for stimulating bone calcification. When combined with varying concentrations of $24R,25\text{-}(OH)_2D_3$ in the presence of PTH, $1\alpha,24R\text{-}(OH)_2D_3$ (0.02 ng/mℓ) gave much the same manner of dose-dependent stimulation as $1\alpha,25\text{-}(OH)_2D_3$ (Figure 2B), although the synthetic vitamin alone with added PTH did not enhance calcification (ratio of 1.03).

D. Ossification Stages Regulated by Their Own Different Combination of Vitamin D_3 Metabolites and PTH

Periosteal ossification of developing bone goes through the following successive processes: (1) proliferation of osteoblasts; (2) elaboration of intercellular organic matrix by these cells; and (3) deposition of bone salt over the matrix by the same cells. The duration of in vitro cultivation of developing bone would roughly correspond to the ossification stages. Therefore, we tried to examine whether three kinds of physiological factors — PTH, $1\alpha,25\text{-}(OH)_2D_3$, and $24R,25\text{-}(OH)_2D_3$ — are always required throughout the ossification process or if different factors are functioning in turn in each stage of the process.

In the control group the bones were cultivated in basal medium with all three factors added, while in experimental groups the pair-mated bones were grown in medium from which a given factor was removed during the first 2 days of cultivation (experiment 1) or the last 4 days (Experiment 2) (Table 3).

When $24R,25\text{-}(OH)_2D_3$ was removed from the medium, the calcification ratio became lower in both Experiments 1 and 2, suggesting that this vitamin D_3 metabolite is essential for all the stages of ossification (Table 3, A and B). Removal of $1\alpha,25\text{-}(OH)_2D_3$ or $1\alpha,24R\text{-}(OH)_2D_3$ from the medium, however, gave decreased calcification only in the later stages of cultivation (Table 3, C and D), probably indicating selective stimulatory action upon rather late stages of the ossification process. On the contrary, PTH was confirmed to be required only in the initial stage of the culture (Table 3, E and F), demonstrating that the hormone is necessary for the earliest stage(s) of ossification. In conclusion, combination of PTH and $24R,25\text{-}(OH)_2D_3$ and of $1\alpha,25\text{-}(OH)_2D_3$ and $24R,25\text{-}(OH)_2D_3$ were found to be prerequisites for the earlier and later stages, respectively, of ossification.

Table 3
THE EFFECT OF FACTOR EXCLUSION ON THE CALCIFICATION OF 9-DAY CHICK EMBRYONIC FEMORA GROWN IN TISSUE CULTURE

Exclusion	Experiment 1 (Mean ± S.E.)	Experiment 2 (Mean ± S.E.)
A. $1\alpha,25\text{-}(OH)_2D_3$ <u>$24R,25\text{-}(OH)_2D_3$</u> PTH	0.81 ± 0.10^a	0.93 ± 0.05
B. $1\alpha,24R\text{-}(OH)_2D_3$ <u>$24R,25\text{-}(OH)_2D_3$</u> PTH	0.68 ± 0.10^b	0.86 ± 0.07^a
C. $1\alpha,25\text{-}(OH)_2D_3$ <u>$24R,25\text{-}(OH)_2D_3$</u> PTH	1.03 ± 0.22	0.82 ± 0.10^a
D. $1\alpha,24R\text{-}(OH)_2D_3$ <u>$24R,25\text{-}(OH)_2D_3$</u> PTH	1.02 ± 0.15	0.75 ± 0.08^b
E. $1\alpha,25\text{-}(OH)_2D_3$ $24R,25\text{-}(OH)_2D_3$ <u>PTH</u>	0.79 ± 0.13	1.15 ± 0.11
F. $1\alpha,24R\text{-}(OH)_2D_3$ $24R,25\text{-}(OH)_2D_3$ <u>PTH</u>	0.72 ± 0.04^b	1.12 ± 0.07

Note: The values represent the ratio (treated bone to control bone) of calcium deposited in the bone after 6 days of cultivation. The substance underlined is excluded from the control medium for the first 2 days in Experiment 1 or for the last 4 days in Experiment 2.

[a] $p < 0.10$.
[b] $p < 0.01$.

E. No Effect of Calcitonin on Bone Calcification In Vitro

Calcitonin is a well-known calcium-regulating hormone which counteracts PTH in bone resorption.[16] As concluding experiments in this work, we examined the effect of calcitonin on bone calcification in vitro (Table 4).

In the first series of experiments, control bones were cultured in basal medium alone. Pair-mated experimental bones were grown first in PTH and calcitonin-containing medium and secondly in $1\alpha,25\text{-}(OH)_2D_3$, $24R,25\text{-}(OH)_2D_3$ and calcitonin-containing medium. No stimulation of calcium deposition during the culture period could be found using either experimental medium (Table 4, Experiment A). These findings indicate that calcitonin can substitute for neither vitamin D_3 metabolites nor PTH in bone calcification in vitro.

In the second experiment, control bones were cultivated in complete medium containing three kinds of calcification-stimulating factors, while treated bones were grown in medium with calcitonin also added. Much the same degree of stimulated calcification occurred in both media, the calcification ratio being 1.02 (Table 4, Experiment B). These results show that calcitonin has no direct effect, stimulatory or inhibitory, upon bone mineralization.

Table 4
EFFECT OF CALCITONIN ON BONE FORMATION

Treatment	Concentration	Calcification ratio[a]	No. of bone pairs
Experiment A			
Control medium:	CEE + BGJb − HW2		
+ PTH	1.0 U/mℓ		
+ Calcitonin	0.5 U/mℓ	1.01 ± 0.07	7
+ 1α,25-(OH)₂D₃	0.02 ng/mℓ		
+ 24R,25- (OH)₂D₃	0.5 ng/mℓ	1.10 ± 0.04	11
+ Calcitonin	0.5 U/mℓ		
Experiment B			
Control medium: CEE + BGJb-HW2 + PTH + 1α,25-(OH)₂D₃ + 24R,25-(OH)₂D₃			
+ Calcitonin	0.5 U/mℓ	1.02 ± 0.09	9

Note: The values represent means ± standard error.

[a]   The ratio of treated femora to pair-mated control bones in terms of calcium level.

IV. DISCUSSION

One of the major unresolved questions concerning the biology of vitamin D is whether increased mineralization elicited by vitamin D in rachitic animals is dependent upon its direct action on bone. On the one hand, 25-OH-D₃, as the first metabolite formed in the liver from vitamin D₃, is further hydroxylated in the kidney to become 1α,25-(OH)₂D₃ and 24R,25-(OH)₂D₃.[17] The former dihydroxy vitamin D₃ metabolite is already well-known to have the strongest biological activity, whereas the latter dihydroxy metabolite has long been considered to be fully inactive.[18] Recently, however, 24R,25-(OH)₂D₃ has been suggested to be necessary for bone formation in living animals.[9-11] In the field of bone biology, on the other hand, bone-forming hormonal factors never have been clarified, whereas at least one bone-resorbing physiological factor has been identified as PTH. In view of the present situation, therefore, it would be a valuable contribution in the field of research to establish an appropriate bone culture system for demonstrating the direct effects on ossification of possible bone formation-controlling factors.

The roller-tube culture method using liquid media for embryonic long bone rudiments was devised[6] and serum was demonstrated to contain some physiological factor or factors which are responsible for inducing periosteal calcification in the presence of chick embryo extract (CEE).[6,7] Concerning cartilaginous growth or elongation of bone, we have recently analyzed the chondrogenic action of PTH and calcitonin[14,19] using a chemically defined medium, BGJb-HW2.[12] In a synthetic medium alone, however, vitamin D₃ metabolites and these hormones could never induce bone calcification.

In this work, therefore, we decided to use a semi-synthetic medium, BGJb-HW2:CEE(9:1), as a basal medium for examining the effects of possible bone calcification-stimulating factors. Moreover, we had two hypotheses: (1) bone-resorbing factors would be bone-forming factors as well, and (2) these factors must be added simultaneously into the medium.

Our findings have clearly demonstrated that these hypotheses are absolutely correct; when simultaneously administered, PTH and vitamin D₃ metabolites can induce bone calcification in vitro. Here it should be noted that PTH would have dual activities for bone formation as well as well-documented bone resorption and that 24R,25-(OH)$_2$D$_3$ would not be fully inactive but very active, at least in bone formation. Considered together, our results can be taken to indicate that bone could be remodelled successfully by simultaneous activation of both osteoclastic and osteogenic processes with a combination of multiple physiological factors.

Our results from the dose-response studies have also confirmed the above findings. These results strongly suggest that the concentration ratios of these three physiological factors are much more important than ever previously considered. From this point of view, therefore, much more detailed studies should be done in the near future. Even at present, however, our results might suggest a possibility that hyperparathyroidism can result in osteosclerosis when the level of 1α,25-(OH)$_2$D$_3$ is normal and that of 24R,25-(OH)$_2$D$_3$ is strikingly high.

The results of the effects of additional analogues suggest that (1) the combination of 1α,25-(OH)$_2$D$_3$ and 24R,25-(OH)$_2$D$_3$ cannot be replaced by a further metabolite having all their structures in a single molecule, and (2) the most active vitamin D$_3$ metabolite, 1α,25-(OH)$_2$D$_3$, can be substituted by a synthetic analogue, 1α,24R-(OH)$_2$D$_3$. Therefore, the biological activities of this synthetic vitamin should be examined in more detail from every aspect of vitamin D action, since it might be considered a possible candidate as a new vitamin D drug.

Our findings concerning the time sequence of factor activities during ossification may be very important, though the experiments are rather preliminary at present. The results strongly suggest that different stages of ossification might require their own sets of controlling factors. Further analyses of the observations may give an insight into their mode of action.

Finally, it was shown that calcitonin has no direct action upon osteogenesis. Phylogenically the hormone is widely distributed even in nonosseous animals, and so our results are not so peculiar.

Considering all these results together, it should be stressed again that our work contains novel, important findings contributing to a better understanding of biology of vitamin D$_3$, bone physiology, and endocrinology of calcium-regulating hormones. Lastly, however, we must add that the role of CEE is not yet clear in this bone culture system. We are now studying the responsible factor(s) or condition(s) in this complex fluid as well as the action mechanism of the calcification-stimulating factors.

REFERENCES

1. **Raisz, L. G.,** Bone resorption in tissue culture. Factors influencing the response to parathyroid hormone, *J. Clin. Invest.*, 44, 103, 1965.
2. **Mahgroub, A. and Sheppard, H.,** Effect of hydroxyvitamin D$_3$ derivative on ^{45}Ca release from rat fetal bones in vitro, *Endocrinology*, 100, 629, 1977.
3. **Stern, P. H. and Raisz, L. G.,** Organ culture of bone, in *Skeletal Research: An Experimental Approach*, Simmons, D. J. and Kunin, A. S., Eds., Academic Press, New York, 1979, 2.
4. **Stern, P. H.,** The vitamins and bone, *Pharmacol. Rev.*, 32, 47, 1980.
5. **Fell, H. B. and Robison, R.,** Growth, development and phosphatase activity of embryonic avian femora and limb-buds cultivated in vitro, *Biochem. J.*, 23, 767, 1929.
6. **Endo, H.,** Ossification in tissue culture. I. Histological development of the femur of chick embryo in various liquid media, *Exp. Cell Res.*, 21, 151, 1960.
7. **Ito, Y., Endo, H., Enomoto, H., Wakabayashi, K., and Takamura, K.,** Ossification in tissue culture. II. Chemical development of the femur of chick embryo in various liquid media, *Exp. Cell Res.*, 31, 119, 1963.
8. **Lieberherr, M., Garabedian, M., Guillozo, H., Bailly, B. M., and Balsan, S.,** Interaction of 24,25-dihydroxyvitamin D$_3$ and parathyroid hormone on bone enzymes in vitro, *Calcif. Tissue Int.* 27, 47, 1979.

9. **Kanis, J. A., Cundy, T., Bartlett, M., Smith, R., Heynen, G., Warner, G. T., and Russell, R. G. G.,** Is 24,25-dihydroxycholecalciferol a calcium regulating hormone in man?, *Br. Med. J.,* 1, 1382, 1978.

10. **Ornoy, A., Goodwin, D., Noff, D., and Edelstein, S.,** 24,25-Dihydroxyvitamin D is a metabolite of vitamin D essential for bone formation, *Nature (London),* 276, 517, 1978.

11. **Bordier, P. and Tun Chot, S.,** Vitamin D metabolite and bone mineralization in man, *J. Clin. Endocrinol. Metab.,* 46, 284, 1978.

12. **Endo, H.,** Cartilage matrix metabolism of chondrocytes changing through their life span and its possible hormonal control, as demonstrated by tissue culture studies (in Japanese with English summary), *Connective Tissue Res.,* 6, 139, 1974.

13. **Endo, H., Kiyoki, M., Kawashima, K., Naruchi, T., and Hashimoto, Y.,** Vitamin D_3 metabolites and PTH synergistically stimulate bone formation of chick embryonic femur in vitro, *Nature (London),* 286, 262, 1980.

14. **Kawashima, K., Iwata, S., and Endo, H.,** Growth stimulative effect of parathyroid hormone, calcitonin and $N^6,O^{2'}$-dibutyryl adenosine 3',5'-cyclic monophosphoric acid on chick embryonic cartilage cultivated in a chemically defined medium, *Endocrinol. Jpn.,* 27, 349, 1980.

15. **Kawashima, H., Hoshina, K., Hashimoto, Y., Takeshita, T., Ishimoto, S., Noguchi, T., Ikekawa, N., Morisaki, M., and Orimo, H.,** Biological activity of 1α,24-dihydroxycholecalciferol: a new synthetic analog of the hormonal form of vitamin D, *FEBS Lett.,* 76, 177, 1977.

16. **Hirsh, P. F. and Munson, P. L.,** Thyrocalcitonin, *Physiol. Rev.,* 49, 548, 1969.

17. **DeLuca, H. F. and Schnoes, H. K.,** Metabolism and mechanism of action of vitamin D, *Ann. Rev. Biochem.,* 45, 631, 1976.

18. **Tanaka, Y., DeLuca, H. F., Kobayashi, Y., Ikekawa, N., and Morisaki, M.,** Biological activity of 24,24-difluoro-25-hydroxy vitamin D_3. Effect of blocking 24-hydroxylation on the functions of vitamin D, *J. Biol. Chem.,* 254, 7163, 1979.

19. **Kawashima, K., Iwata, S., and Endo, H.,** Selective activation of diaphyseal chondrocytes by parathyroid hormone, calcitonin, and $N^6,O^{2'}$-dibutyryl adenosine 3',5'-cyclic monophosphoric acid in proteoglycan synthesis of chick embryonic femur cultivated in vitro, *Endocrinol. Jpn.,* 27, 357, 1980.

20. **Connerty, H. V. and Briggs, A. R.,** Determination of serum Ca by means of *o*-cresolphthalein complexone, *Am. J. Clin. Pathol.,* 45, 290, 1966.

Chapter 3

VITAMIN C AND INFECTIOUS DISEASE: A REVIEW OF THE LITERATURE AND THE RESULTS OF A RANDOMIZED, DOUBLE-BLIND, PROSPECTIVE STUDY OVER 8 YEARS

Maxine Briggs

TABLE OF CONTENTS

I. REVIEW OF THE LITERATURE

A. In Vitro Studies

Any pure chemical substance may have antimicrobial activity, and most do. The mechanisms involved, however, are very diverse. All compounds with significant water solubility at physiological temperature are likely to inhibit microbial growth when present at high concentrations due to simple osmotic effects. Those compounds which are also acids or bases may significantly alter pH, which will also be injurious to microbial survival. Finally, a minority of substances will be deleterious to microbes due to specific effects on their metabolism, even when used at physiological pH and osmolality.

In reviewing the enormous published literature dealing with ascorbic acid (AA) and infections, an obvious starting point is the in vitro effects on microorganisms. A summary of reported actions of AA on cells, microorganisms and microbial toxins in vitro is given in Table 1. In many cases it is not possible to determine whether antagonistic effects of AA are due to specific biochemical actions, or to failure to control pH and/or osmotic effects.

There is no doubt, however, that AA inhibits the growth of some bacteria and inactivates certain viruses and bacterial toxins following incubation at physiological temperature and pH at low concentrations (10^{-4} to $10^{-6} M$). There are some conflicting results in the published studies, while several reports suggest that inhibitory action of AA is due to some metabolite or oxidation product. An alterative hypothesis would be that AA reduces some disulfide bridges of key proteins and so changes their tertiary structure and function.

B. Animal Studies

Numerous studies have been published of infections in laboratory animal models treated with AA (Table 2). The most frequently used species has been the guinea pig, raised on a scorbutic diet. Experiments have also been made with mice and monkeys. The usual finding has been that many infections nonspecifically reduce tissue levels of AA and increase dietary AA requirements in species lacking the ability to synthesize vitamin C.

There are some indications for immunological effects of AA. The development of antibodies is reduced by a lack of AA, while the vitamin may stimulate interferon production and phageocytosis.

With specific regard to influenza, while antibody production was increased in mice receiving extra AA, there was no effect on mortality from influenza or the extent of lung lesions.

Some of these animal studies provide evidence for an unfavorable adaptation to high AA intakes in guinea pigs, in that animals which have undergone such adaptation appear more susceptible to the development of scorbutic symptoms when suddenly returned to a normal or deficient AA intake.

C. High-Dose Vitamin C and the Common Cold

The belief that a large intake of AA can prevent or reduce the incidence of colds, or can have curative effects on a preexisting cold, dates back to at least the early 1940s, with several studies being published at that time (see Table 3). Considerable public attention was given to this view by Linus Pauling in a popular book first published in 1971 (later extended and reprinted in 1976). Another serious paperback on the same topic was published by Irwin Stone in 1974, while there have been many copyists, often of doubtful merit and ability.

Evaluation of published studies on high-dose AA and the common cold is complicated for a variety of reasons, some of which are

1.   There is no single disease entity that can be identified with the common cold. A wide range of rhinoviruses are involved in cold epidemics, together with viruses in other

Table 1
IN VITRO EFFECTS OF AA ON CELLS, MICROORGANISMS, AND TOXINS

Author	Microorganism or cells	Reported effects of AA	Ref.
Jungeblut	Poliomyelitis virus	Sodium ascorbate pH 6.6—6.8 inactivated polio virus	188
Jungeblut and Zwemer	Diphtheria toxin	AA at pH 6.6—6.8 inactivated diphtheria toxin	192
Grooten and Bezssonoff	Various bacilli	AA has antibacterial action in cultures	156
Holden and Resnick	Herpesvirus	AA in buffers pH 5—8 does not inactivate herpesvirus	176
Lominski	Bacteriophage	Inactivated by AA at pH 7	238
Amato	Rabies virus Tetanus toxin	Inactivation of fixed virus attenuation and inactivation of toxin	8
Boissevain and Spillane	Tubercule bacilli	Human TB bacilli in artificial media were inhibited by 0.001% AA	61
Holden and Molloy	Herpesvirus	AA at pH 6 inactivated herpesvirus	175
Jungeblut	Tetanus toxin	AA at pH 6 inactivated tetanus toxin	189, 190
Kligler and Bernkopf	Vaccinia virus	Inactivated by AA	217
Kligler et al.	Diphtheria toxin	Inactivated by AA	219
Schulze and Hecht	Diphtheria toxoid Tetanus toxin	Inactivated by AA	327
Kligler et al.	Tetanus toxin	Inactivated by AA	218
Kodama and Kojima	Staphlococcal hemolysins	Not inactivated by AA but growth of staphylococci inhibited	222
Martin	Tobacco mosaic virus	Inactivated by AA undergoing oxidation, but not by either fully oxidized or fully reduced AA; accelerated inactivation in presence of copper ions	248
Myrvik et al.	Mycobacteria	Inhibited by AA oxidation products	265
Ericsson and Lundebeck	*Escherichia coli*	AA has antibacterial action in culture, but is much weaker in body fluids	128
Walker et al.	Cold viruses in fibroblast cell cultures	No antiviral effect of AA treatment	373
Miller	Gram-negative bacteria	AA + H_2O_2 is bacteriocidal to a variety of bacteria, perhaps by effects on cell walls	258
Cooper et al.	Human neutrophils Rabbit macrophages	AA stimulates hexose monophosphate shunt pathway	90
Murata and Kitagawa Murata et al.	Bacteriophages	Inactivated by AA in vitro	263 262
Rawal et al.	*Pseudomonas aeruginosa*	Inhibited by AA in vitro Antimicrobials act synergistically with AA	303
Murata	Bacteriophages	A variety of bacteriophages is inactivated by AA in vitro	264

Table 1 (continued)
IN VITRO EFFECTS OF AA ON CELLS, MICROORGANISMS, AND TOXINS

Author	Microorganism or cells	Reported effects of AA	Ref.
Schwerdt and Schwerdt	Rhinovirus in W1-38 cell cultures	AA + glutathione suppressed multicyclic, but not single-cycle, growth of the virus	330

Note: This table lists information in chronological sequence.

Table 2
ANIMAL STUDIES OF AA AND INFECTIONS

Author	Infections	Reported effects	Ref.
Harde and Philippe	Diphtheria	AA reduces effects of toxin on guinea pigs	160
Jungeblut and Zwemer	Diphtheria toxin	AA treatment helps to protect guinea pigs from diphtheria intoxication	192
Grootten and Bezssonoff	Diphtheria	Simultaneous administration of AA with diphtheria toxin to guinea pigs reduces the toxic effects	156
Harris et al.	Tuberculosis or diphtheria toxin	Guinea pigs infected with TB or diphtheria toxin had low adrenal levels of AA; liver AA was reduced only in chronic TB infection	162
Holden and Molloy	Herpesvirus	AA treatment did not influence infection with dermal herpesvirus	175
Jungeblut	Poliomyelitis	Monkeys received polio virus intracerebrally, together with various doses of AA; small supplements were more protective than large doses	189
Sigal and King	Diphtheria toxin	Disturbance of glucose tolerance in guinea pigs was less if high-dose AA was also given	338
Jungeblut	Poliomyelitis	AA supplements without effect when large doses of virus given intranasally; smaller doses less effective	191
Sabin	Poliomyelitis	AA had no effect on experimental disease in monkeys receiving virus by nasal installation; deficient animals died more frequently of acute infection	316
Cottingham and Mills	Pneumococcus	Phageocytosis stimulated in infected guinea pigs by AA	91
Russell and Callaway	Trypan blue deposition	Heavier than normal deposition of dye in liver and kidneys of scorbutic guinea pigs	315

Table 2 (continued)
ANIMAL STUDIES OF AA AND INFECTIONS

Author	Infections	Reported effects	Ref.
Sadun et al.	*Entamoeba histolytica*	High infectivity, mortality, and severity of symptoms were seen in AA-deficient guinea pigs	318
Willis	Liver disease	Scorbutic guinea pigs develop hepatic fatty degeneration and necrosis that are reversible by AA, but not by cystine and/or choline	377
Gordonoff	Miscellaneous	Guinea pigs receiving high doses of AA die earlier than those receiving physiological amounts when both groups switched to a scorbutic diet	150
Cochrane	Miscellaneous	Guinea pigs exposed to high doses AA *in utero* showed an induced dependency and susceptibility to scurvy	88
Walker et al.	Influenza A virus	Mortality rate and extent of lung lesions in infected mice were not influenced by i.p. 300 mg/kg AA	373
Rawal et al.	*Pseudomonas aeruginosa*	Infected mice are cured by AA alone; the dose required is lower if given in combination with erythromycin	303
Siegel	Murine leukemia virus	Mice given ad lib AA in drinking water had higher circulating interferon than controls	337
Babes et al.	Influenza	Mice receiving high-dose AA (0.5 mg per 14 g animal) given influenza antigen developed significantly lower levels of HAI antibodies than controls	37
Leibovitz and Siegel	Anaphylaxis	Tissue levels of AA in mice are decreased by immunization, with or without anaphylaxis; histamine levels were lower in lung and spleen in AA treated mice following anaphylactic death	232

Note: This table lists information in chronological sequence.

classifications. The viruses are inconsistent, in that regular mutations occur to strains of changed virulence. Some signs and symptoms of a cold are allergic and possibly unrelated to primary virus infections being caused by environmental allergens such as pollens and dust.

2. For relatively trivial illnesses, which are rarely life-threatening, the most appropriate study of any potential prophylactic/therapeutic agent is a placebo-controlled, double-blind, randomized trial, possibly with a cross-over. Some published studies have used this design, but others have been nonrandomized, or only single-blind, or even not blinded. Given the difficulties of diagnosis, of quantifying the diverse signs and

Author	Year	Randomized	Blinding	Placebo	AA	Placebo	AA	Placebo	AA	Country	Ref.
Anderson et al.	1974	Yes	Double	578	583	6.00	6.03	3.18	3.28	Canada	17
Anderson et al.	1972	Yes	Double	411	407	5.92	5.51	4.18	3.96	Canada	16
	1973										14
Baird et al.[a]	1979	Yes	Single	112	118 120	?	?	?	?	U.K.	39
Barnes	1961	No	None	16	22	8.25	6.00	6.5	2.0	U.S.	41
Bessel-Lorck[a]	1959	No	Double	26	20	?	?	?	?	Germany	52
Briggs[b]	1973	Yes	Double	28	33	0.46	0.56	3.2	3.3	Australia	68
Carr et al.	1981	Yes	Double	95	95	1.48	1.58	6.39	5.18	Australia	74
Carr et al. (Identical twins)	1974	Yes	Double	142	153	N.S.D.[c]	3.24	N.S.D.[c]	3.5	U.K.	74
Charleston and Clegg	1972	No	None	43	47	6.45		4.2		U.K.	81
Clegg[a]	1974										84
Clegg and MacDonald[a]	1975	Yes	Double	70	67	3.60	3.50	7.6	7.2	U.K.	85
Coulehan et al.[a]	1976	Yes	Double	129	133	2.63	2.55	5.8	5.5	U.S.	95
Coulehan et al.	1974	No	Double	320	321	0.46	0.40	5.92	4.71	U.S.	97
Cowan and Diehl	1950	Yes	Single	76	77	2.72	2.76	5.1	5.6	U.S.	99
Cowan et al.	1942	No	Single	120	227	2.40	2.40	1.00	1.70	U.S.	100
		No	Single	194	233	2.20	1.90	1.60	1.90	U.S.	
Dahlberg et al.	1944	No	None	1266	1259	0.51	0.40	?	?	Scandinavia	101
Elliott	1944	No	None	90	60	1.28	1.13	4.0	3.2	U.K.	123
Elwood et al.	1973	No	Double	37	33	?	?	?	?	U.S.	126
	1976	Yes	Double			5.33	5.01	7.27	5.81	U.K.	125
	1977			Total =	1082						
Franz et al.	1956	Yes	Double	45	44	1.33	1.27	?	?	U.S.	139
Glazebrook and Thomson	1942	No	None	1100	335	0.52	0.43	4.9	2.5	U.K.	147
Karlowski et al.	1975	Yes	Double (partial)	89	101	1.81	1.69	6.30	6.80	U.S.	195

Wilson et al.										
Woolstone	1954	No	Single	?	?	?	?	?	?	U.K.

Note: This table lists information in chronological sequence.

[a] Results scaled up from short treatment period.
[b] Placebo was 50 mg AA daily.
[c] N.S.D. = not significantly different.

symptoms, and of the variability of colds with time and between individuals, only randomized, double-blind studies can possibly be considered to provide any important contribution to possible effects of high-dose AA.

3.   Unlike other drugs used against infectious diseases, AA is also a nutrient, though being used at nonphysiological doses. Nevertheless, in double-blind studies, especially where there is no cross-over, it is essential to be certain that nutrient intake is adequate in both groups, so that high-dose AA is not being used merely to correct a deficiency of vitamin C that could have been corrected with a much lower dose. Very few of the published studies have determined either the dietary vitamin C intake of their volunteers, or any laboratory index of vitamin C status (such as leukocyte concentrations or 24-hr urinary output).

The general impression from these studies is that high-dose AA has little or no effect in preventing or treating common colds. It is possible that the incidence of colds may be slightly reduced by prophylactic AA, while the duration of colds that do occur may be slightly shorter. Any such effect, however, is very small, and is not consistently reported from all trials.

Many medical and scientific writers have commented upon these publications, as well as providing additional information on possible effects of AA on colds. Several have also published monographs on potential adverse effects. These numerous commentaries are summarized in Table 4. Aside from a few enthusiasts, such as Pauling, Stone, and Wilson, the overall impression is that the international medico-scientific community is unimpressed with the evidence relating high-dose AA to either the prevention or treatment of the common cold.

Several authors have commented that even if there is a small positive effect, the potential benefit does not counterbalance the unknown risks of prolonged AA intake at high doses. The comment has been made that any other drug showing such weak therapeutic properties would be dropped from further development by the pharmaceutical industry.

Despite these comments, a number of formulations containing AA have been developed specifically as anti-cold preparations (Table 5).

D. Vitamin C in Other Infections

There is less literature on the use of vitamin C in other infections. Table 6 summarizes published findings on tuberculosis and includes some relevant information on animal models. There seems no doubt that AA requirements are increased during active tuberculosis and that tissue levels are reduced. Deficient animals seem more susceptible to infection and the same may be true of humans.

There is a lack of evidence that high-dose AA is of value in the treatment of tuberculosis or its complications. The same is true of pneumonia (Table 7), though hypovitaminosis C may predispose to more serious disease.

Of the four published studies on whooping cough (Table 8), there may be tissue depletion of AA during the infection, but high-dose AA does not improve the course of the disease, nor reduce the incidence of complications.

Evidence relating AA to various forms of hepatitis (Table 9) is rather sparse, but a good controlled study suggests that high-dose AA is unable to reduce the incidence of posttransfusion hepatitis.

For poliomyelitis the studies are very limited and inadequately controlled (Table 10). The same is true for leprosy (Table 11).

A range of miscellaneous conditions is listed in Table 12, but most are uncontrolled, or inadequately controlled, and refer to very small series.

Numerous comments have been published on the world literature relating to infectious

Table 4
COMMENTS ON AA AND COLDS

Author	Comment	Ref.
Otani	Uncontrolled study of AA therapy in prophylaxis and treatment of respiratory infections	271
Widenbauer	Possible adverse kidney changes with high-dose AA	375
Kuttner	A multivitamin supplement containing AA had no effect on the incidence of respiratory infections in rheumatic children	226
Van Alyea	1.0 g AA daily useful in treatment of early rhinosinusitis	365
Dujardin	Uncontrolled study of 10 g AA daily in 14 patients with infections	115
Markwell	Uncontrolled studies suggest that high doses of AA may prevent (350 mg daily) or be used to alleviate symptoms (750 mg daily) of colds	245
Neuweiler	High doses of AA reduce fertility and during pregnancy increase fetal death	267
Patterson	High doses of dehydro-AA are diabetogenic in rats and act synergistically with alloxan; much less effects are seen with high-dose AA	274
Mouriguand and Edel	High doses of AA may influence fertility and pregnancy	260
Macon	Patients with colds were given either an AA-bio-flavonoid-aspirin or an aspirin-phenacetein-caffeine mixture; on the second day of medication, 74% of 62 patients receiving the first and 51% of 59 patients receiving the second, reported complete or substantial relief of symptoms	244
Tebrock et al.	Approximately 2000 patients with upper respiratory infections received 0.2 g AA daily, sometimes with 1.0 g bioflavonoid, or placebo; none of the treatments significantly influenced the course of common colds	356
Miegl	Uncontrolled study of 247 patients with variety of upper respiratory tract acute infections who received high doses of AA	255
Barnes	U.S. high school basketball players received a daily multivitamin supplement containing 200 mg AA for 8 weeks; the average incidence of colds in 23 treated students was 1.3 days for girls and 1.6 days for boys, compared to 6.9 days in 13 untreated controls	41
Pena et al.	High-dose AA increases the urinary excretion of uric acid	289
Samborskaja	Large doses of AA terminated pregnancy in guinea pigs	320
Querton	Uncontrolled study of a formulation containing AA on symptoms of colds and influenza	301
Mehnert et al.	Young, nondiabetic subjects (12) received 1.5 g AA daily for 6 weeks; there was no change in glucose tolerance, or in urinary glucose measured specifically	252
Sambroskaja and Ferdman	6 g AA daily for 3 days terminated early pregnancy in 16 of 20 women, apparently by increasing estrogen production	321

Table 4 (continued)
COMMENTS ON AA AND COLDS

Author	Comment	Ref.
Anonymous	There is no conclusive evidence that, in the absence of severe AA depletion, vitamin C has any effect on the incidence, course, or duration of colds	33
Kimbarowski and Mokrow	A urinary colored precipitation reaction was carried out on 214 patients with influenza (viral grippe); treatment with 300 mg AA daily for 2 weeks reduced the number of positive reactions	202
Tyrrell	3 g AA daily is unable to prevent the appearance of colds due to 5 different viruses	361
Walker et al.	Volunteers received placebo or 3 g AA daily for 3 days before challenge with cold viruses; there was no difference between the groups on the incidence or severity of colds	373
Wilson	Uncompleted studies in Dublin suggest that prophylactic use of high-dose AA against colds is not totally negative	379
Abbott et al.	Large daily doses of vitamin C did not alter the duration or symptoms in patients with colds	2
Regnier	Review of literature and account of uncontrolled trial of high-dose AA in treatment of colds	304
Wilson and Loh	Plasma AA values are altered in school children during colds; prophylactic administration of AA reduces the intensity of symptoms	395
Anonymous	The daily use of large doses of vitamin C for prophylaxis against colds cannot be recommended	23, 24
Ketz	AA does not exert any detectable favorable effect as a prophylactic or therapeutic agent against colds and infections	199
Kobza et al.	A group of ironmakers and a group of coalminers received 200 mg AA daily for 3—4, months; as compared to unsupplemented workers, sick leave was reduced, especially for influenza and other respiratory illnesses	221
Stott and Taylor-Robinson	The number, duration, and severity of colds is the same in subjects taking high-dose AA as in untreated controls	347
Vodopija	Large doses of AA have been claimed to reduce the incidence of colds, but published studies found no protection against cold viruses	371
Pauling	Arguments that the human daily dietary requirements for AA have been seriously underestimated	286, 288
Anonymous	Studies in Canada by Anderson suggest a slight beneficial effect of vitamin C supplement on colds; tissue saturation can be achieved with 120 mg daily, which may be easily obtained by selecting foods rich in vitamin C	32, 34
Beaton and Whalen	The use of vitamin C to prevent colds is unsubstantiated	46
Goldsmith	"There is no scientific evidence supporting the use of large doses of AA in preventing or treating the common cold"	148
Goldstein	Woman developed sickle cell crisis with self-administered high-dose AA, but not otherwise	149
Hoffer	Psychiatric patients given high doses of AA spontaneously report low incidence of colds; this does not occur with other high-dose vitamins	174

Table 4 (continued)
COMMENTS ON AA AND COLDS

Author	Comment	Ref.
Pauling	Reviews of published studies on AA and colds, which concludes that high-dose AA reduces the incidence, severity, and morbidity	278, 279, 288
Rhead and Schrauzer	High-dose AA may predispose to scurvy when stopped and usual dietary intake is resumed	306
Rosenthal	Case report of shortening of prothrombin time in warfarin-treated woman who took high doses of AA	312
Wilson	The efficacy of AA against colds requires critical interpretation of AA metabolism during infections	389
Blum	Value of high-dose AA is unproved	58
Anderson et al.	The frequency and duration of colds was not statistically significantly different between a vitamin C supplemented group and a placebo group, though more of the supplemented group remained free of illness	16
Isaksson	No convincing evidence that high doses of AA are of medical importance; prevention of the common cold is still unanswered due to poor design of published studies	183
Ketz	Present experience of the prophylactic use of AA is insufficient to justify its large-scale application	200
Lamy and Rotkovitz	No scientific evidence that AA has an alleviating or curative effect on colds	228
Masek et al.	AA supplements to miners reduced days lost due to sickness, especially from respiratory disease	250
Poser	Briefly discusses man taking 4 g AA daily who experienced loss of visual acuity, which resolved spontaneously, and continued AA supplement for 13 years without problem	299
Preshaw	Effects of AA in prevention and treatment of common cold are still in doubt	300
Riccitelli	AA may suppress inflammation associated with colds and render virus penetration of cells more difficult	307
Cheraskin et al.	Survey of dental practitioners for vitamin C intake and respiratory signs and symptoms found a significant negative correlation	83
Hume and Weyers	Leukocyte AA fell significantly within 24 hr of onset of cold symptoms	181
Lewin	Suggests that AA may interfere with organization of viruses by lowering surface tension	234
Sakakura et al.	AA had no effect on susceptibility to induced rhinovirus infection, or on mucociliary transport	319
Schrauzer and Rhead	Abrupt cessation of high-dose AA may lead to unexpected deficiency symptoms due to conditioning	326
Schwartz et al.	3 g AA daily was of no prophylactic benefit against induced rhinovirus infection in 11 men, compared with 10 placebo-treated controls	328
Spero and Anderson	It is doubted that high-dose AA will prove to be of any great clinical value, while little is known of possible harmful effects	342
Wilson et al.	Administration of 0.2 g to 0.5 g AA during colds benefits symptoms more in girls than boys	400

Table 4 (continued)
COMMENTS ON AA AND COLDS

Author	Comment	Ref.
Wilson and Loh	At 0.5 g daily AA reduces the severity and duration of colds in girls, but not in boys; the frequency of toxic and catarrhal symptoms is altered	397, 398
American Academy of Pediatrics Committee on Drugs, 1974	There is insufficient evidence that high doses of vitamin C is either safe or efficacious in the prevention or treatment of the common cold and it should not be used for this purpose until such data are available	
Anderson et al.	The optimum daily dose of vitamin C is less than 250 mg, except possibly at the time of illness when a larger intake may be beneficial	17
Anderson	Large scale Canadian study found more subjects free of illness in vitamin group than placebo controls ($p < 0.05$), but mean days of illness were not significantly different	12
Bouhuys	Vitamin C may have an antihistaminic effect on the human respiratory tract.	63
Martin and Lines	High-dose AA interferes with Benedict's test for urinary glucose, but not "Clinistix"; there is no sound evidence that AA supplements prevent colds	247
Masek et al.	Extra supplies of AA improve phagocytic activity of white cells, rate of collagen formation, and amino acid metabolism; these effects may relate to the prevention and treatment of infections, such as the common cold	249
Williams	Small changes in urinary oxalate are important in the development of calcium nephrolithiasis	376
Wilson	Tissue saturation with AA protects against colds and other infections	388
Wilson and Greene	2 g AA during colds causes a significant rise in female leukocyte AA, but not in males	391
Wilson and Loh	During the common cold there is a sex-linked difference in AA metabolism; a single 1 g dose of AA raised AA in blood in both sexes, but it was maintained for 2 hr longer in females	394
Anderson et al.	Vitamin C supplemented groups experienced less severe winter illnesses than placebo groups	13
Chalmers	A detailed review on published studies suggests that any effect of high-dose vitamin C in preventing or ameliorating colds is very minor	79
Greene and Wilson	Aspirin enhances in vivo uptake of AA by leukocytes in subjects with colds, but has much less effect after recovery	153
Schneider	High-dose AA has some effect on the severity and duration of cold symptoms	325
Seifter	Brief comment on new findings with high-dose AA	333
Wilson	Cold symptoms are associated with altered AA metabolism; utilization of AA increases during viral and allergic colds for implementation of tissue defense mechanisms	385—387
Wilson	AA supplements limit the severity and duration of the common cold syndrome	383, 384

Table 4 (continued)
COMMENTS ON AA AND COLDS

Author	Comment	Ref.
Editorial	International double-blind studies have failed to demonstrate that high-dose AA is effective in preventing colds	120
Hejda et al.	Administration of 100 mg AA daily to miners reduced the incidence of respiratory disease; best results gathered from influenza vaccination together with supplementary AA	169
Wilson et al.	In subjects with colds, 2 g AA increased plasma AA in both sexes, but leukocyte AA in girls only; high tissue concentrations of AA were associated with low toxic and catarrhal symptoms	399
Anderson	Regular daily supplements of vitamin C with extra dosage at the time of illness reduced disability due to colds and other winter illness, but was without effect on frequency of illness	11
Cook and Monsen	Increase in iron absorption from a semi-synthetic diet was proportional to added AA over range 25—1000 mg	89
Hughes	In man, tissue AA saturation is attainable with 100—150 mg daily and there are no compelling reasons to use megadoses	179
Miller et al.	Monozygotic twin pairs (44) received 0.5—1.0 g AA daily on a double-blind basis with co-twins receiving a placebo; treated females, but not males, had less severe respiratory tract infections; treated males grew more than untreated pairs; there were no effects on serum cholesterol, but urinary amino acid excretion decreased specifically	256, 257
Tyrrell	Some trials have shown a weak effect of AA against colds, though properly designed trials have usually shown little or none	360
Vallance	Leukocyte AA levels are positively correlated with IgG and IgM	364
Epstein and Masek	A study of compliance in college students taking vitamin C q.i.d	127
Kent	Brief review of published studies concludes that the evidence that vitamin C is effective in treating respiratory disease is relatively slim	197, 198
Taft and Fieldhouse	Insufficient acceptable evidence that AA is an efficacious and safe treatment for colds; further well-designed trials are needed	351
Wilson and Greene	Elevation of leukocyte AA by high-dose AA during colds is eliminated by aspirin	392
Anderson	Large daily doses of vitamin C will not prevent or cure the majority of colds; increased intake at the time of illness may reduce severity	9
Chang and Syndman	Review of antiviral agents briefly mentions AA and comments that controlled studies have failed to show any benefit of large doses to reduce the incidence or duration of colds	80
Coulehan	Controlled double-blind studies do not support the contention that high doses of AA prevent colds or abort colds	92—94

Table 4 (continued)
COMMENTS ON AA AND COLDS

Author	Comment	Ref.
Davies et al.	Subjects infected with common cold viruses on a controlled daily intake of AA showed a significant reduction in urinary AA and diketogluconate	104
Deni	Uncritical commentary on Pauling's hypothesis	109
Editorial	Vitamin C does not seem to decrease the incidence of colds and winter illness	122
Elsborg	Lack of evidence that AA is effective against colds	124
Bee	Unsupported claim that 10—15 g AA daily is necessary to relieve cold symptoms	48
Gorman	Concern that placebo tablets often do not exactly match AA tablets in color and taste	151
Hill and Kamath	Children with diarrhea staining napkins pink were found to be taking an AA supplement containing sorbitol	171
Shadrin et al.	Uncontrolled study of AA in influenza	335
Vilter	Megadoses of AA for prevention of colds are ineffective, and in a few cases, dangerous	369
Wilson	AA is valuable in preventing the symptoms of respiratory diseases by inhibiting excessive inflammatory response to exogenous proteins	382
Rinke	0.5 g AA reduced bronchospasm in group of patients with exercise-induced asthma	308

Note: This table lists information in chronological sequence.

Table 5
ANTICOLD FORMULAS CONTAINING AA

Author	Formulation	Ref.
Franz and Heyl	65 mg AA 333 mg Maringin (flavonoid)	139
Macon	50 mg AA 100 mg Citrus flavonate glycoside 292 mg Acetylsalicylic acid	244
Tebrock et al.	50 mg AA 250 mg Lemon bioflavonoid complex 497	356
Querton	40 mg AA 1.25 mg Phenylephrine 100 mg Acetylaminophenol 0.25 mg Dimethylpyridine maleate 15 mg Trihydroxyrutoside	301
Allen and Das Gupta	250 mg AA 50 mg Methapyrilene HCl 5 mg Phenylephrine HCl	7
Gromova et al.	150 mg AA 250 mg Acetylsalicylic acid 50 mg Calcium lactate 10 mg Rutin 10 mg Dimidrol	155

Note: This table lists information in chronological sequence.

Table 6
AA AND TUBERCULOSIS (TB)

Author	Findings	Ref.
McConkey and Smith	Of two groups of guinea pigs given TB sputum, 26 of 37 fed an AA-deficient diet developed intestinal lesions, compared to only 2 of 35 receiving AA supplements	241
Abbasy et al.	24-hr excretion of AA significantly lower (mean 9 mg) in 24 patients with active TB than in 46 with quiescent TB (mean 19 mg): diets unsupplemented with AA	1
Greene et al.	Scorbutic guinea pigs with TB had a shorter survival than nonscorbutic, but TB-infected animals allowed to develop scurvy had similar illness to controls	152
Hasselbach	Uncontrolled studies on various AA formulations in patients with active pulmonary TB	163
Heise and Martin	Vitamin C requirements are increased in human active TB	167
Boissevain and Spillane	Treatment of guinea pigs with 25 mg AA daily s.c. did not protect against TB infection	61
Martin and Heise	AA treatment (up to 400 mg daily) appeared to improve prognosis in pulmonary TB, but study inadequately controlled	246
Radford et al.	TB patients receiving supplementary AA showed better results than controls for hemoglobin, RBC count, lymphocytes, monocyte-lymphocyte ratio, neutrophil-lymphocyte ratio, and albumin-globulin ratio	302
Birkhaug	Hypovitaminosis-C renders the guinea pigs more vulnerable to progressive TB; high AA doses (10 mg p.o. daily) significantly reduce invasive lesions and development of generalized TB	57
Jetter and Bumbalo	Urinary AA was significantly reduced in 37 children with active TB	186
Ruskin	Uncontrolled study of AA calcium salt in acute rhinitis	314
Bakhsh	20 patients received 500 mg vitamin C i.m. daily for 4 days and 150—200 mg p.o. daily for 6 weeks; therapeutic effects were most marked in secondary anemia; the E.S.R. was reduced in many cases; cough was reduced, but expectoration, temperature, and lung signs were little altered	40
Heise and Steenken	High-dose AA does not influence the course of TB in infected guinea pigs, nor the tuberculin sensitivity	166
Josewich	Urinary AA was low in patients with TB; an uncontrolled study suggests that selective patients improve when given 100—150 mg AA daily	187
Erwin et al.	Patients were saturated with AA, but it was of no value in the treatment of pulmonary TB or its complications	130
Kaplan and Zonnis	Patients with chronic pulmonary TB are often deficient in vitamin C; treatment of 101 patients for 6 months with 200 mg AA daily had no effect on course of their disease	194

Table 6 (continued)
AA AND TUBERCULOSIS (TB)

Author	Findings	Ref.
Bogen et al.	Hypovitaminosis-C increases the incidence of serious mucous membrane complications of TB; patients receiving vitamin supplements containing 150 mg AA daily showed visible improvements to ulcerated mucous membrane lesions	59
Sweany et al.	Depletion of AA levels occurs with progressive TB; intake should be increased to 0.2 g daily depending on the duration, severity, and activity of the disease	348
Steinbach and Klein	Tuberculous guinea pigs given AA showed better tolerance to large doses of tuberculin and retarded progress of TB	344
Pijoan and Sedlacek	Patients with TB had low plasma AA concentrations and required higher than normal dietary AA intake	296
Rudra and Roy	Improved blood picture in 14 patients with pulmonary TB given 0.25 g AA daily	313
Babbar	Patients receiving 200 mg vitamin C daily for 2.5 months showed improved hemoglobin, erythrocyte count, and staff cells in Schilling count, but no significant change in lymphocytes or E.S.R.	36
Charpy	Uncontrolled trial of 15 g AA daily in combination with supplementary vitamin D_2	82
Getz et al.	Over a 7-year period, 28 men from a group of 1100, initially disease-free, developed X-ray signs of TB; all cases of active disease had poor AA status prior to TB development	146
Myrvik et al.	AA supplements improved the tuberculostatic activity of sterile urines tested in culture	265
Boyden and Andersen	Supplementary AA improved the survival times of guinea pigs infected with virulent TB	66

Note: This table lists information in chronological sequence.

Table 7
AA AND PNEUMONIA

Author	Reported effects	Ref.
Gander and Niederberger	Incidence, severity, and mortality from pneumonia related to inadequate AA status	141
Kienast	Uncontrolled study of AA supplements in 20 cases of pneumonia, briefly reported	201
Biilmann	Uncontrolled study of 400—500 mg AA daily p.o. with M + B693 in small series	54
Stein	Mortality from pneumonia can be reduced by the use of 0.1—0.5 g AA daily, with thiamin and corticosteroids, as an adjunct to chemotherapy and serum	343
Zlydnikov and Chepuk	Uncontrolled study of intensive therapy including AA in patients with pneumonia during an influenza epidemic	406

Note: This table lists information in chronological sequence.

Table 8
AA AND WHOOPING COUGH

Author	Results	Ref.
Ormerod et al.	Patients with whooping cough show low urinary AA; saturation with AA supplements decreased severity of symptoms in uncontrolled study	270
Plate	Uncontrolled study of AA therapy in 5 children	298
Gairdner	21 cases treated with 200 mg AA daily, reducing step-wise to 100 mg daily; no significant difference between AA supplements and unsupplemented children	140
Pfeiffer	There was no significant difference in the improvement of complications in 36 patients receiving high-dose AA than in 35 who did not receive supplements	294

Note: This table lists information in chronological sequence.

Table 9
AA AND HEPATITIS

Author	Condition	Results	Ref.
Baur and Staub	Hepatitis epidemica	Daily infusions of 10 g AA for 5 days improved recovery in 11 patients compared to others not receiving AA	45
Kirchmair	Hepatitis	Children with epidemic hepatitis received i.v. infusion containing high-dose AA; as compared with previous epidemics, AA shortened the illness, while liver enlargement was less	204, 205
Calleja and Brooks	Acute hepatitis	A 44-year-old man received 5 g AA daily i.v. for 24 days; ascites resolved and LFT became normal; serial liver biopsies showed disappearance of neutrophilic infiltration	71
Knodell et al.	Posttransfusion hepatitis	Randomized double-blind study of 800 mg AA q.d.s. vs. placebo in 175 cardiac surgery patients; incidence and course of hepatitis was same in both groups	220

Note: This table lists information in chronological sequence.

Table 10
AA AND POLIOMYELITIS

Author	Results	Ref.
Baur	Patients receiving high-dose vitamin C (10—20 g i.v. or p.o.) showed better improvement than un-supplemented patients: small numbers and inade-quately controlled	44
Gsell and Kalt	Doses of AA (5—25 g p.o. or 0.02—0.51 g/kg i.v.) were given during a polio epidemic; severity of symptoms was decreased with increasing AA doses	157
Greer	Uncontrolled study on high-dose AA in patients with acute polio	154

Note: This table lists information in chronological sequence.

Table 11
AA AND LEPROSY

Author	Report	Ref.
Bechelli	Patients received 50—100 mg AA daily i.m. for treatment of leprous reaction; there was improve-ment in about 50% of cases	47
Gatti et al.	Improvement in two lepers receiving i.v. AA: oral AA used to control leprotic septicemia	143
Ugarizza	Uncontrolled study of 0.4 g AA daily to treat le-prous septicemia	363
Floch and Sureau	Uncontrolled study of 2—4 g AA daily	138

Note: This table lists information in chronological sequence.

Table 12
MISCELLANEOUS CONDITIONS

Author	Condition	Reported findings	Ref.
Spengler	Liver cirrhosis	AA has diuretic effect	341
Farah	Enteric fever	Uncontrolled study of 15 patients treated with adrenal cortex extract plus i.v. AA	131
Takahasi	Dysentery	Uncontrolled study of AA against Shiga-bacillus and Komagome B-bacillus	352
Mick	Brucellosis	Large doses of AA (up to 4 g daily) im-proved some patients in uncontrolled study of 12 subjects	254
Dugal	Cold temperature tolerance	Large doses of AA improve tolerance of animals exposed to low temperatures	116
Dymock et al.	Gastroduodenal disorders	Patients had low amounts of buffy coat AA	118
Hindson	Prickly heat	Double-blind study found that AA (15 mg/kg) was significantly better than a placebo	172
Klenner	Various virus infections	Massive doses of AA given in encepha-litis and other conditions; lack of ade-quate controls	212

Table 12 (continued)
MISCELLANEOUS CONDITIONS

Author	Condition	Reported findings	Ref.
Hume	Anticoagulant therapy	1 g AA daily does not influence effects of warfarin	180
Cameron and Pauling	Carcinomatosis	Proposal to inhibit tumor growth by AA enhancement of glycosoaminoglycan synthesis to restrain proliferation	72
Taylor et al.	Pressure sores	Prospective double-blind study in 20 surgical patients suggests that 0.5 g AA daily accelerates healing of pressure sores	355
Gerson	Ileitis	Dietary intake and leukocyte and serum AA lower in ileitis patents than controls; serum, but not leukocyte, response to AA supplement lower in ileitis	145
Zlydnikov	Influenza	Uncontrolled study of remantadin plus	405
Zlydinkov et al.		AA	407
Hornig	Smoking	New data on AA turnover suggests that an adult man requires about 100 mg daily; the requirement in smokers is about 140 mg daily	177

Note: This table lists information in chronological sequence.

disease and AA (Table 13). The major, widespread opinion is that many infections non-specifically reduce tissue AA concentrations, and so increase requirements, but high-dose AA is of little or no value in the prophylaxis or specific treatment of human infectious diseases.

E. Wound Healing

There is general agreement on the basis of limited studies (Table 14) that wound healing is impaired in AA deficiency, but stimulated by AA treatment.

F. Conclusions

Despite the very large literature on AA and infectious disease, many publications are anecdotal case histories, or poorly controlled studies. While AA is able to inactivate some infectious microorganisms and their toxins in vitro, this does not necessarily happen in vivo.

The balance of evidence strongly suggests that many different infections deplete tissue stores of AA and so increase daily requirements. Similarly, some infections occur more readily in AA-depleted subjects. High-dose AA, however, is only weakly prophylactic, if at all, while it is of little or no use in the treatment of human infections.

II. CLINICAL INVESTIGATION: AN 8-YEAR, PROSPECTIVE, DOUBLE-BLIND STUDY OF HIGH- VS. LOW-DOSE AA SUPPLEMENTATION FOR THE PREVENTION OF COMMON COLDS

A. Introduction

The hypothesis put forward by proponents of the ''orthomolecular'' school of therapeutics with regard to vitamin C and the common cold may be summarized as follows:

1. The usually recommended daily dietary allowance for vitamin C is sufficient to prevent scurvy, but insufficient to protect against the common cold (and related infections)

Table 13
AA AND INFECTIONS: REVIEWS

Author	Principal conclusions	Ref.
Faulkner and Taylor	Patients with various infections have low serum AA and large doses are required to restore normal values	132
Perla and Marmorston	Avitaminosis C is associated with reduced natural resistance to infection; excess AA increases the resistance of guinea pigs to diphtheria and TB toxins, anaphylaxis, and trypanosome infections; infectious diseases increase the oxidation of AA, and so increase requirements	290, 291
Thaddea and Hoffmeister	Infectious disease reduces the concentration of AA in blood, CSF, and urine	357
Abt and Farmer	Low AA intake lowers resistance; some infections increase AA requirements; AA is not a specific therapy for any infection	3
Bourne	AA improves the resistance of the body to infective agents by effects on complement, acquired immunity, and possible enhanced inactivation of viruses	64
McCormick	The incidence and case-fatality of many infectious diseases has decreased dramatically over the past century; an important factor increased availability of fruits and consequent higher AA intake	243
McCormick	Intensive AA therapy is an important adjunct in the treatment of many infectious diseases, including TB and diphtheria	242
Dalton	Uncontrolled studies on the use of 2 g AA + B vitamins i.v. in 6 patients with viral diseases	103
Heinrich	Brief review on possible benefits and hazards of high-dose AA	165
Catalano	Comments on the role of vitamin C lack and Sjögren's syndrome	76
Perrault and Day	Infections reduce AA concentration in leukocytes so that supplementary AA may improve treatment	292
Dykes and Meier	No clear reproducible pattern of efficacy has come from studies of high-dose AA to prevent or treat the common cold	117
Ganguly et al.	Orange juice reduces symptoms of viral respiratory infections; AA may enhance functions of phagocytic cells	142
Thomas and Holt	AA is involved in the migration and phagocytosis by macrophages and leukocytes, as well as in delayed hypersensitivity; effects on antibody production and complement are minimal	359
Stone	It is hypothesized that high-dose AA stimulates interferon synthesis	346
Woolfe et al.	Deficiency of AA produces only minimal changes in gingival health, while AA supplements have produced conflicting results in gingival inflammation	401

Note: This table lists information in chronological sequence.

Table 14
AA AND WOUND HEALING

Author	Reported effects	Ref.
Bartlett et al.	Healing and tensile strength are decreased in human AA deficiency	42
Bourne	Lack of AA impairs wound healing	65
Schwartz	AA promotes wound healing	329

Note: This table lists information in chronological sequence.

2. The adult daily intake of vitamin C to prevent the common cold is about 1000 mg
3. Should symptoms of a common cold develop, these may be eliminated or reduced by increasing the daily intake of vitamin C to about 4000 mg while symptoms persist

Evidence relevant to this hypothesis has been itemized in detail in Section I. A major difficulty facing all clinical studies on the efficacy of high-dose vitamin C in preventing the common cold, or reducing the severity of its symptoms, is that sub-clinical deficiency of vitamin C appears to predispose to infections. It is clear, therefore, that unless the vitamin C intake of any control group is at least the recommended daily allowance, the high-dose treated group may be compared not with subjects receiving the usually accepted daily intake, but with a sub-normal group. This would clearly introduce a bias into the studies.

With this point in mind, the present study was designed in which a high dose of vitamin C was compared, not with an inert placebo, but with a preparation containing 50 mg AA daily. In this way there could be no doubt that the controls were receiving the usually accepted adequate intake.

B. Methods and Materials

The basis of the study was to randomize carefully screened volunteers between two identical products, one containing 50 mg and the other 1000 mg L-ascorbic acid. The obvious solution appeared to be to pack these weights of AA into identical opaque gelatin capsules (dark brown) and to make up the weight difference with some other white crystalline powder, of similar acidic taste, but lacking vitamin C activity. Citric acid was selected (950 mg to each 50 mg AA capsule).

In order to ensure this mixture with citric acid did not either interfere with the absorption or pharmacokinetics of AA, studies were made in 10 volunteers who received capsules of 50 mg AA alone and 50 mg AA + 950 mg citric acid on a double-blind, cross-over basis. Measurements were made of plasma AA at 1-hr intervals for 8 hr, and a 24-hr urine was collected and examined for AA and metabolites.

There were no significant differences between the preparations with or without citric acid for all of these parameters.

Vitamin C capsules stored at room temperature in bottles with silica gel bags showed similar mild deterioration (98 ± 2% S.D. stability) after 100 days, irrespective of the presence of citric acid.

Capsules were packed in brown bottles (100 capsules per bottle). Each bottle was labelled with an adhesive patch with a code number, depending upon the date of manufacture and packaging. Bottles of capsules of the two types were separately crated (each clearly marked A or B).

The criteria for inclusion in the study are set out in Table 15 which summarizes the complete protocol of the study. Volunteers were assigned to product A or B by the use of

Table 15
PROTOCOL FOR THE COMPARATIVE STUDY OF HIGH- AND LOW-DOSE VITAMIN C

1. Adults (either sex) in full-time employment (aged 18 + years)
2. Cigarette intake <10 daily
3. No oral contraceptive use
4. No regular use of other drugs
5. Not pregnant or lactating
6. At least 6 months postpartum
7. No other vitamin supplements (fruit juices were not excluded)
8. No contraindications to high-dose vitamin C
9. No history of nasopharyngeal pathology
10. No history of asthma, chronic bronchitis, pneumonia, etc.
11. Vegetarians excluded
12. Judged to be intelligent and well-motivated
13. Capsules for 100 days supplied: bottle to be returned for residual count and replacement; further return at 6 months
14. Only subjects returning at 3 or 6 months included in study: total subjects divided for analysis by total known duration of treatment
15. No current illness: at least 3 months since previous illness (no chronic disease)
16. Volunteers assigned to 1 of the 2 products using random number tables
17. One product contained 1000 mg AA the other 50 mg AA plus 950 mg citric acid (both in identical brown gelatin capsules): products identified by code numbers on label
18. Instructions were to take 1 capsule daily and increase to 4 capsules daily at first signs of a cold and continue 4 while symptoms persisted
19. Code broken only after subject dropped from study (3 or 6 months completed)
20. Re-entry into the study was not allowed
21. Subjects asked to rate severity of any cold by 7 symptoms, each graded severe, moderate, or mild (see Table 2); record card supplied
22. The symptoms were explained to each subject with care until the interviewer was satisfied that the subject understood how to grade each symptom accurately
23. Subjects supplied written informed consent
24. Selected subjects supplied 24-hr dietary histories (by recall), 5 mℓ blood by venepuncture (for ascorbate analysis), and 24-hr urine collections (for ascorbate and oxalate analysis)

random number tables and neither the physician nor the volunteer was aware of the composition of the capsules prescribed.

Clinical criteria for the classification of colds were based on those of Abbott et al. (1968) and are shown in Table 16. The record card completed by each subject is shown in Table 17.

Dietary histories were obtained from some subjects (using the 24-hr recall method) during the return visits after 3 or 6 months. These subjects also brought with them a 24-hr urine collection and supplied 5 mℓ of nonfasting blood (collected by venepuncture of the antecubital vein). Urine was analyzed for total ascorbate (using the method of Roe and Kuether, 1943) and oxalate (using the enzymic method of Costello et al., 1976). Heparinized blood was separated into plasma, platelets, and leukocytes, which were separately analyzed for total AA (methods of Albanese et al., 1975 and Briggs, 1973).

C. Results and Discussion

In all, 528 persons took part in this study (160 men and 368 women). Distribution of both sexes between the two doses of vitamin C was close (see Table 18). Low-dose AA was received by 263 persons (186 women and 77 men), while the 1000 mg dose was taken by 265 persons (182 women and 83 men).

Of the total group (528 persons), 237 completed only a single 3-month course (45%),

Table 16
CLINICAL CRITERIA FOR SYMPTOMS[2]

Symptom	Grade[a]	Criteria
Sore throat	1	Red and inflamed throat
		Sore all day long
		Pain on swallowing
		Requiring analgesics
	2	Less red and inflamed throat
		Sore mainly in mornings and on eating
		Requiring only occasional analgesics
	3	Sore only for short time first thing in mornings
Stuffy nose	1	Both nostrils completely blocked, resulting in mouth breathing
		Blockage present more often than it is not
		Nostrils cannot be cleared by blowing
	2	Intermittent obstruction of nostrils
		Obstruction may be one-sided
		Nostrils sometimes cleared by blowing
	3	Nostrils not blocked and nose-breathing possible
		Blowing necessary to keep nostrils clear
Sneezing	1	More or less continuous throughout the day
	2	Bouts on and off throughout the day
	3	Occasional (usually early mornings only)
Watery nasal discharge	1	All day long
		Continuous use of handkerchief
	2	Bouts on and off throughout the day
		Intermittent use of handkerchief
	3	Short time only (usually mornings)
		Occasional use of handkerchief
Purulent nasal discharge	1	Thick and purulent
	2	Less thick and less purulent
	3	Muco-purulent
Headache	1	Continuous analgesics required
	2	Occasional analgesics required
	3	Ache present but no analgesics required
		Not inconveniencing in any way
Aching back and limbs	1	Continuous analgesics required
	2	Occasional analgesics required
	3	Aches present, but analgesics not required
		Not inconveniencing in any way

[a] Grades: 1-severe; 2-moderate; 3-mild

while 291 (55%) completed the double course of 6 months. The sex distributions were as follows:

50 mg	Men completing 3 months	= 29 } 77	(38% total)
	Men completing 6 months	= 48	(62% total)
	Women completing 3 months	= 89 } 174	(51% total)
	Women completing 6 months	= 85	(49% total)
1000 mg	Men completing 3 months	= 34 } 83	(41% total)
	Men completing 6 months	= 49	(59% total)
	Women completing 3 months	= 85 } 182	(47% total)
	Women completing 6 months	= 97	(53% total)
Both doses	Men completing 3 months	= 118 } 263	(45% total)
	Men completing 6 months	= 145	(55% total)
	Women completing 3 months	= 174 } 368	(47% total)
	Women completing 6 months	= 194	(53% total)
Both doses	M, F completing 3 months	= 237 } 528	(45% total)
	M, F completing 6 months	= 291	(55% total)

Table 17
RECORD CARD

DATE OF NEXT APPOINTMENT: _____
(Please bring this card with you)
NAME: _____

Write the date for each day with cold symptoms and evaluate yourself for the severity of each of the seven symptoms (check the appropriate boxes). Try to fill in the record daily, rather than rely on memory. Leave a blank where a symptom is not present.[a]

Dates of colds	Sore throat			Stuffy nose			Sneezing			Watery nasal discharge			Purulent nasal discharge			Headache			Aching back and limbs		
	1	2	3	1	2	3	1	2	3	1	2	3	1	2	3	1	2	3	1	2	3

[a]   Scored 4 in the analysis.

Generally speaking, significantly more men entering the study completed the 6-month course than women, who had a higher drop-out rate at the 3-month stage.

Table 19 lists the annual and cumulative days of treatment with the two preparations. Over the 8-year period the total number of treatment days for both doses combined was 25,079; 12,349 days with the 50 mg dose and 12,730 days with 1000 mg dose. Men used the 50 mg dose for 3781 days (30.6% of total for that dose), while women used 50 mg for 8568 days (69.4%). For the 1000 mg dose, men used this for 4220 days (33.1% of total for that dose) and women for 8510 days (66.8%).

A detailed breakdown of reported cold symptoms is given in Table 20. Over the period of study there was a total of 246 colds, of which 121 (49.2%) occurred in the 50 mg dose group and 125 (50.8%) in the 1000 mg dose group. This is not significantly different. The number of days on which cold symptoms occurred was 792, of which 402 (50.8%) were with the 50 mg dose and 390 (49.2%) with 1000 mg. Again, these are not significantly different.

The mean duration of cold symptoms was 3.3 days with 50 mg and 3.1 days with 100 mg.

These data are presented in Table 21 as percentages of the total groups (by year and cumulative). Only occasional individuals reported more than one cold during the trial period. Figures 1 to 5 graphically present the data presented in the above tables.

Estimates of the severity of colds are given in Table 22, where the six major symptoms were separately assessed by subjects during each cold episode. There are clearly major differences in certain symptoms during particular years. For example, aching back and limbs were much less frequent in 1977 through 1981 than in 1974 and 1975. Similarly, stuffy nose was less severe in 1977 and 1979 than in most other years. Sore throat and sneezing, however, were relatively constant throughout the 8-year study period.

When the average scores for each symptom are compared between the 50 mg and 1000 mg treatment groups, it is immediately apparent that differences between the scores for stuffy nose, headache, and aching back are very small and statistically insignificant.

For the other symptoms (sore throat, sneezing, and watery nasal discharge), the 1000 mg

Table 18
ANNUAL AND CUMULATIVE TREATMENTS

		Subjects						Cumulative					
		Men		Women		Total		Men		Women		Total	
Year[a]		50	1000	50	1000	50	1000	50	1000	50	1000	50	1000
1974	T	11	13	17	20	28	33	—	—	—	—	28	33
	3	5	6	7	11	12	14	—	—	—	—	12	17
	6	6	7	10	9	16	16	—	—	—	—	16	16
1975	T	12	10	22	29	34	39	23	23	39	49	62	72
	3	6	4	11	14	17	18	11	10	18	25	29	35
	6	6	6	11	15	17	21	12	13	21	24	33	37
1976	T	16	18	43	26	59	44	39	41	82	75	121	116
	3	6	9	21	12	27	21	17	19	39	37	56	56
	6	10	9	22	14	32	23	22	22	43	38	65	60
1977	T	9	9	30	31	39	40	48	50	112	106	160	156
	3	3	4	12	13	15	17	20	23	51	50	71	73
	6	6	5	18	18	24	23	28	27	61	56	89	83
1978	T	8	8	22	20	30	28	56	58	134	126	190	184
	3	3	4	10	9	13	13	23	27	61	59	84	86
	6	5	4	12	11	17	15	33	31	73	67	106	98
1979	T	11	8	29	29	40	37	67	66	163	155	230	221
	3	3	2	14	11	17	13	26	29	75	71	101	100
	6	8	6	15	18	23	24	41	37	88	84	129	121
1980	T	8	6	19	17	27	23	75	72	182	172	257	244
	3	2	3	12	10	14	13	28	32	87	81	115	113
	6	6	3	7	7	13	10	47	40	95	91	142	131
1981	T	2	11	4	10	6	21	77	83	186	182	263	265
	3	1	2	2	4	3	6	29	34	89	85	118	119
	6	1	9	2	6	3	15	48	49	97	97	145	146

[a] T = total; 3 = 3 months; 6 = 6 months.

Table 19
ANNUAL AND CUMULATIVE DAYS OF TREATMENT

	Men		Women		Total		50 +
Year	50	1,000	50	1,000	50	1,000	1,000
1974	512	695	818	879	1,330	1,574	2,904
	—	—	—	—	—	—	—
1975	541	609	997	1,333	1,538	1,942	3,480
	1,053	1,304	1,815	2,212	2,868	3,516	6,384
1976	784	815	1,959	1,218	2,743	2,033	4,776
	1,837	2,119	3,774	3,430	5,611	5,549	11,160
1977	453	426	1,451	1,490	1,904	1,916	3,820
	2,290	2,545	5,225	4,920	7,515	7,465	14,980
1978	395	368	1,029	939	1,424	1,307	2,731
	2,685	2,913	6,254	5,859	8,939	8,772	17,711
1979	575	426	1,332	1,440	1,907	1,866	3,773
	3,260	3,339	7,586	7,299	10,846	10,638	21,484
1980	428	272	796	728	1,224	1,000	2,224
	3,688	3,611	8,382	8,027	12,070	11,638	23,708
1981	93	609	186	483	279	1,092	1,371
	3,781	4,220	8,568	8,510	12,349	12,730	25,079

Table 20
SUBJECTS AND DAYS WITH REPORTED COLD SYMPTOMS

Year	Subjects								Totals				Mean cold duration	
	Men				Women									
	50		1000		50		1000		50		1000		50	1000
	S	D	S	D	S	D	S	D	S	D	S	D		
1974	5	16	8	27	8	25	11	36	13	41	19	63	3.2	3.3
1975	5	21	6	24	12	49	14	56	17	70	20	80	4.1	4.0
	(10	37)ª	(14	51)	(20	74)	(25	92)	(30	111)	(39	143)	(3.7	3.7)
1976	8	20	6	17	13	46	13	38	21	66	19	55	3.0	2.9
	(18	57)	(20	68)	(33	120)	(38	130)	(51	177)	(58	198)	(3.5	3.4)
1977	4	7	3	7	12	23	15	29	16	30	18	36	0.9	2.0
	(22	64)	(23	75)	(45	143)	(53	159)	(67	207)	(76	294)	(3.1	3.1)
1978	5	21	4	26	7	28	7	19	12	49	11	45	3.9	4.1
	(27	85)	(27	101)	(52	171)	(60	178)	(79	256)	(87	279)	(3.2	3.2)
1979	6	28	4	21	15	62	9	39	21	90	13	60	4.3	4.6
	(33	113)	(31	122)	(67	233)	(69	217)	(100	346)	(100	339)	(3.5	3.4)
1980	4	14	4	8	13	37	10	28	17	51	14	36	3.0	2.6
	(37	127)	(35	130)	(80	270)	(79	245)	(117	397)	(114	375)	(3.4	3.3)
1981	1	2	5	6	3	3	6	9	4	5	11	15	1.2	1.3
	(38	129)	(40	136)	(83	273)	(85	254)	(121	402)	(125	390)	(3.3	3.1)

Note: S = subjects; D = days.

ª Numbers in parenthesis indicate cumulative totals.

Table 21
PERCENTAGE SUBJECTS AND DAYS WITH COLDS: ANNUAL AND CUMULATIVE

| | Men | | | | Women | | | | Total | | | |
| | 50 | | 1000 | | 50 | | 1000 | | 50 | | 100 | |
Year	S	D	S	D	S	D	S	D	S	D	S	D
1974	45	3.1	61	3.9	47	3.1	55	4.1	46	3.1	58	4.0
	—	—	—	—	—	—	—	—	—	—	—	—
1975	41	3.9	60	3.9	54	4.9	48	4.2	50	4.5	51	4.1
	(43	3.5)[a]	(61	3.9)	(51	4.1)	(51	4.2)	(48	3.9)	(54	4.1)
1976	50	2.5	33	2.1	30	2.3	50	3.1	36	2.4	43	2.7
	(46	3.1)	(51	3.2)	(40	3.2)	(51	3.8)	(42	3.1)	(50	3.6)
1977	44	1.5	33	1.6	40	1.6	48	1.9	41	1.6	45	1.9
	(46	2.8)	(48	2.9)	(40	2.7)	(50	3.2)	(42	2.7)	(49	3.1)
1978	62	5.3	37	7.1	54	2.7	75	2.0	53	3.4	64	3.4
	(48	3.2)	(40	3.5)	(34	2.7)	(42	3.0)	(35	2.9)	(41	3.2)
1979	54	4.9	50	4.9	52	4.6	31	2.7	52	4.7	35	3.2
	(49	3.5)	(47	3.6)	(41	3.1)	(44	3.0)	(43	3.2)	(45	3.2)
1980	50	3.3	66	2.9	68	4.6	59	3.8	63	4.2	61	3.6
	(49	3.4)	(49	3.6)	(44	3.2)	(46	3.1)	(45	3.3)	(47	3.2)
1981	50	2.1	45	1.0	75	1.6	60	1.9	66	1.8	52	1.4
	(49	3.4)	(48	3.2)	(45	3.2)	(47	3.0)	(46	3.3)	(47	3.1)

Note: S = subjects; D = days.

[a] Numbers in parenthesis indicate cumulative totals.

treatment group is slightly better than the 50 mg group to the extent of approximately $+6\%$ in each case. By the χ^2 test, however, this difference fails to reach statistical significance.

It should be stressed that if years are considered in isolation, occasional statistically significant differences between the two treatment groups can be found (e.g., stuffy nose in 1980, sore throat in 1975, aching back and limbs in 1978). These differences, however, disappear when the complete 8-year treatments are pooled. As the common cold viruses are subject to frequent mutations, it might be argued that vitamin C is effective against certain strains, but not others. While this possibility cannot be discounted, it would be almost impossible to investigate by clinical trials other than on a massive scale with detailed identification of strains by viral immunology and electron microscopy.

The distribution of AA between blood plasma, platelets, and leukocytes for 34 persons taking 50 mg vitamin C and for 36 taking 1000 mg is shown in Table 23. While plasma levels were slightly higher in the 1000 mg group, the difference is not statistically significant. AA concentrations in platelets and leukocytes were also not significantly different between the two treatment groups, nor between the sexes, either within a group or between groups.

Finally, Table 24 gives the estimated dietary vitamin C intake of 143 persons in the 50 mg dose group and 151 in the 1000 mg group. As expected, the mean urinary AA excretion per 24 hr was highly significantly greater for the 1000 mg group than for the 50 mg group ($p < 0.001$). While 24-hr urinary oxalate values overlapped between the two groups, the mean value in the 1000 mg group was significantly higher than in the 50 mg group ($p < 0.01$).

D. Conclusions

1.   A total of 528 carefully selected individuals were randomized between either 50 mg

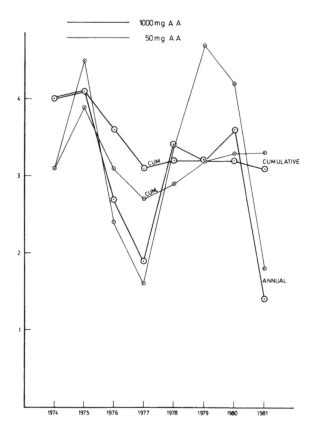

FIGURE 1.   Total percentage days with colds

or 1000 mg vitamin C daily for 3 or 6 months. There were no significant differences between the two groups for the number of colds, their severity, or duration.

2.   The concentrations of AA in blood plasma, platelets, and leukocytes were not significantly different between the two groups, but those receiving 1000 mg daily excreted much more AA in 24-hr urine specimens. The high-dose group also excreted significantly more oxalic acid, though there was considerable overlap between the two groups.

3.   The present study suggests that a daily large intake of vitamin C (1000 mg increasing to 4000 mg during a cold) is no more helpful in reducing the incidence, severity, or duration of colds than a much smaller dose (50 mg increasing to 200 mg during a cold). Although the increase in urinary oxalate with the high dose is small, this may increase the risk of urinary stones in predisposed individuals. High doses of vitamin C should not, therefore, be used for mass self-medication.

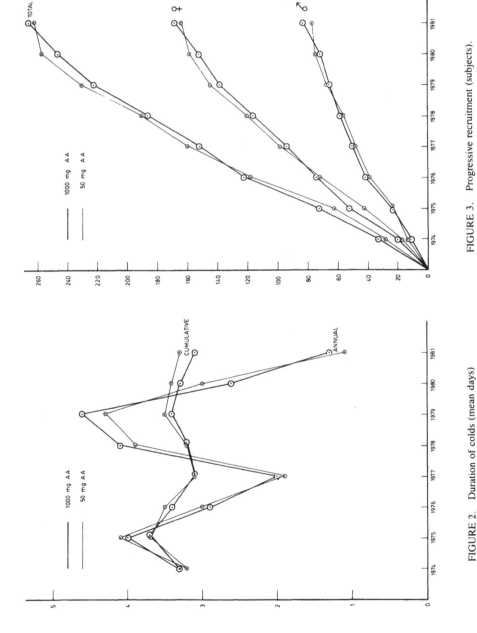

FIGURE 3. Progressive recruitment (subjects).

FIGURE 2. Duration of colds (mean days)

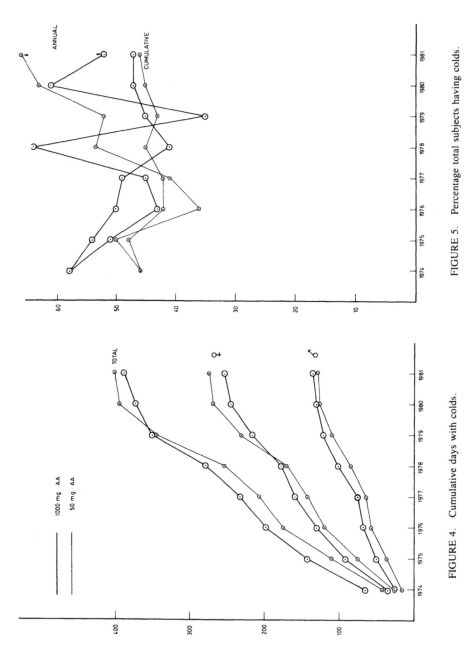

FIGURE 5. Percentage total subjects having colds.

FIGURE 4. Cumulative days with colds.

Table 22

SEVERITY OF COLDS: SYMPTOM SCORES[a]

Year	Dose (mg)	Sore throat		Stuffy nose		Sneezing		Watery nasal discharge		Headache		Aching back	
		Total	Average/cold	Total	Average/cold	Total	Average/cold	Total	Average/cold	Total	Average/cold	Total	Average/cold
1974	50	23	1.80	55	4.21	21	1.62	76	5.81	116	8.91	95	7.31
	1000	31	1.61	76	3.98	31	1.65	128	6.72	168	8.87	141	7.40
1975	50	24	1.40	62	3.62	35	2.05	65	3.81	152	8.93	119	7.01
	1000	41	2.05	80	4.02	42	2.10	81	4.06	180	8.65	147	7.02
1976	50	34	1.61	62	2.95	64	3.06	60	2.86	178	8.51	188	8.95
	1000	30	1.59	57	3.01	49	2.57	72	3.79	163	8.57	168	8.85
1977	50	17	1.09	75	4.70	20	1.28	32	2.01	131	8.16	154	9.65
	1000	21	1.15	83	4.62	34	1.90	46	2.57	158	8.75	175	9.72
1978	50	20	1.68	37	3.05	25	2.06	70	5.86	116	9.63	103	8.50
	1000	21	1.91	31	2.85	24	2.15	69	6.31	106	9.61	106	9.68
1979	50	26	1.26	89	4.26	39	1.86	86	4.11	199	9.49	187	8.91
	1000	17	1.31	63	4.82	25	1.90	65	5.01	111	8.51	129	9.89
1980	50	33	1.98	71	4.16	30	1.76	66	3.86	150	8.83	169	9.95
	1000	22	1.61	87	6.21	28	2.00	28	2.00	139	9.90	125	8.93
1981	50	8	2.01	9	2.38	8	2.00	4	1.00	35	8.75	39	9.65
	1000	20	1.85	24	2.16	33	3.00	12	1.11	91	8.30	102	9.29
Total	50	185	1.53	490	4.05	242	2.00	459	3.79	1077	8.90	1053	8.70
	1000	203	1.62	501	4.01	266	2.13	501	4.01	1116	8.93	1093	8.74

[a] Severity scores for all cold episodes in each group. It should be noted that severity of each symptom is *inversely* proportional to the score.

Table 23
AA IN BLOOD (MEAN VALUES ± S.D.)

Group (mg)	Sex	No.	Blood fractions		
			Plasma mg/ℓ	Platelets mg/g wet wt.	Leukocytes µg/10⁸
50	♂	15	11 ± 3	0.27 ± 0.10	34 ± 6
	♀	19	12 ± 3	0.29 ± 0.11	33 ± 5
	Both	34	11 ± 3	0.28 ± 0.10	33 ± 5
1000	♂	14	12 ± 3	0.29 ± 0.09	32 ± 6
	♀	21	13 ± 3	0.27 ± 0.11	34 ± 5
	Both	36	13 ± 3	0.28 ± 0.10	33 ± 5

Table 24
MEAN URINARY EXCRETIONS (± S.D.)

Group (mg)	No.	Food AA[b]	Urinary oxalate and ascorbate[a] (mg/24 hr)	
			Oxalate Mean ± S.D. (range)	Ascorbate Mean ± S.D. (range)
50	143	85 ± 5	46 ± 11 (22—105)	28 ± 9 (19—51)
1000	151	78 ± 8	89 ± 15 (41—151)	705 ± 35 (612—855)

[a]   Subjects provided urine specimens after 3 months treatment: oxalate-rich foods were excluded for 7 days prior to urine collection.
[b]   Estimated from recall and dietary tables.

REFERENCES

1. **Abbasy, M. A.,** Vitamin C and juvenile rheumatism, *J. Soc. Chem. Ind.,* 55, 841, 1936.
2. **Abbott, P.,** Seventy-seven others. Ineffectiveness of Vitamin C in treating Coryza, *Practitioner,* 200, 442, 1968.
3. **Abt, A. F. and Farmer, C. J.,** Vitamin C: pharmacology and therapeutics, *JAMA,* 111, 1555, 1938.
4. **Albanese, P.,** Treatment of respiratory infections with high doses of Vitamin C, *El Dia Med.,* 19, 1738, 1947.
5. **Albrecht, E.,** Vitamin C as an adjuvant in the therapy of lung tuberculosis, *Med. Klin. (Munich),* 34, 972, 1938.
6. **Alexander, H.,** Vitamin C and tuberculosis, *Dtsch. Tuberk. Bl.,* 14, 125, 1940.
7. **Allen, A. E., and Das Gupta, V.,** Quantitative determination of phenylephrine hydrochloride, ascorbic acid and methapyrilene hydrochloride in an anticold formula, *Am. J. Hosp. Pharmacol.,* 30, 77, 1973.
8. **Amato, G.,** Azione dell'acido ascorbico sul virus fisso della rabbia e sulla tossina tetanica, *G. Batteriol. Virol. Immunol. (Torino),* 19, 843, 1937.
9. **Anderson, T. W.,** Vitamin C: cure for the common cold? *Am. Pharmacol.,* 19, 46, 1979.
10. **Anderson, T. W.,** Large scale trials of vitamin C, *Ann. N.Y. Acad. Sci.,* 258, 498, 1975.
11. **Anderson, T. W.,** Large scale studies with vitamin C, *Acta Vitaminol. Enzymol.,* 31, 43, 1977.
12. **Anderson, T. W.,** Large scale trials of vitamin C in the prevention and treatment of "colds", *Acta Vitaminol. Enzymol.,* 28, 99, 1974.
13. **Anderson, T. W., Beaton, G. H., Corey, P. N., and Spero, L.,** Winter illness and vitamin C: the effect of relatively low doses, *Can. Med. Assoc. J.,* 112, 823, 1975.
14. **Anderson, T.W., Reid, D. B., and Beaton, G. H.,** Vitamin C and the common cold, *Can. Med. Assoc. J.,* 108, 133, 1973.

15. **Anderson, T. W., Reid, D. B., and Beaton, G.H.,** Vitamin C and serum cholesterol, *Lancet,* 2, 876, 1972.
16. **Anderson, T. W., Reid, D. B., and Beaton, G. H.,** Vitamin C and the common cold: a double blind trial, *Can. Med. Assoc. J.,* 107, 503, 1972.
17. **Anderson, T. W., Suranyi, G., and Beaton, G. H.,** The effect on winter illness of large doses of vitamin C, *Can. Med. Assoc. J.,* 111, 31, 1974.
18. **Anon.,** Ascorbic acid and common colds, *Br. Med. J.,* 3, 311, 1973.
19. **Anon.,** Good health — the simple way, *Nurs. Times,* 72, 1587, 1976.
20. **Anon.,** Vitamin C and the common cold, *Ceylon Med. J.,* 16, 1971.
21. **Anon.,** Vitamin C and the common cold, *Nutr. Rev.,* 32, 39, 1974.
22. **Anon.,** Vitamin C and the common cold, *Nutr. Rev.,* 31, 303, 1973.
23. **Anon.,** Vitamin C and the common cold, *Med. Lett. Drugs Ther.,* 12, 105, 1970.
24. **Anon.,** Vitamin C — were the trials well controlled and are large doses safe? *Med. Lett. Drugs Ther.,* 13, 46, 1971.
25. **Anon.,** Megavitamin therapy: evidence still sketchy, *Can. Med. Assoc. J.,* 119, 81, 1978.
26. **Anon.,** Ineffectiveness of vitamin C in treating Coryza, *Practitioner,* 200, 442, 1968.
27. **Anon.,** The common cold — vitamin C, antibiotics, 1-antigen B, etc., *Drug Ther. Bull.,* 5, 35, 1967.
28. **Anon.,** Vitamin C may enhance healing of caustic corneal burns, *JAMA,* 243, 623, 1980.
29. **Anon.,** Ascorbic acid — immunological effects and hazards, *Lancet,* 1, 308, 1979.
30. **Anon.,** Primate studies indicate that subclinical and acute vitamin C deficiency may lead to periodontal disease, *JAMA,* 246, 730, 1981.
31. **Anon.,** Vitamin C for pateints with leprosy, *JAMA,* 233, 188, 1975.
32. **Anon.,** Is vitamin C really good for colds? *Consumer Rep.,* February, 68, 1976.
33. **Anon.,** Ascorbic acid and the common cold, *Nutr. Rev.,* 24, 228, 1967.
34. **Anon.,** Vitamin C, Linus Pauling and the common cold, *Consumer Rep.,* February, 113, 1971.
35. **Anon.,** Vitamin C and the common cold, *Med. J. Aust.,* 1, 1361, 1971.
36. **Babbar, I. J.,** Therapeutic effect of ascorbic acid in tuberculosis, *Indian Med. Gaz.,* 83, 409, 1948.
37. **Babes, V. T., Militaru, M., and Lenkei, R.,** Effects of vitamin A and C overdose on formation of influenza HAI antibodies in the mouse, *Virologie,* 27, 63, 1976.
38. **Baetgen, D.,** Results of the treatment of epidemic hepatitis in children with high doses of ascorbic acid in the years 1957—58, *Med. Monatsschr.,* 15, 30, 1961.
39. **Baird, I. M., Hughes, R. E., Wilson, H. K., Davies, J. E., and Howard, A. N.,** The effects of ascorbic acid and flavonoids on the occurrence of symptoms normally associated with the common cold, *Am. J. Clin. Nutr.,* 32, 1686, 1979.
40. **Bakhsh, I. and Rabbani, M.,** Vitamin C in pulmonary tuberculosis, *Indian Med. Gaz.,* 74, 274, 1939.
41. **Barnes, F. E., Jr.,** Vitamin supplements and the incidence of colds in high school basketball players, *N. C. Med. J.,* 22, 22, 1961.
42. **Bartlett, M. K., Jones, C. M., and Ryan, A. E.,** Vitamin C and wound healing. II. Ascorbic acid content and tensile strength of healing wounds in human beings, *N. Engl. J. Med.,* 226, 474, 1942.
43. **Bartley, W., Krebs, H. A., O'Brien, J. R. P.,** Medical Research Council Special Report Series No. 280, Her Majesty's Stationery Office, London, 1953.
44. **Baur, H.,** Poliomyelitis therapy with ascorbic acid, *Helv. Med. Acta,* 19, 470, 1952.
45. **Baur, H. and Staub, H.,** Therapy of hepatitis with ascorbic acid infusions, *Schweiz. Med. Wochenschr.,* 84, 595, 1954.
46. **Beaton, G. H., and Whalen, S.,** Vitamin C and the common cold, *Can. Med. Assoc. J.,* 105, 355, 1971.
47. **Bechelli, L. M.,** Vitamin C therapy of the lepra reaction, *Rev. Bras. Leprol.,* 7, 251, 1939.
48. **Bee, D. M.,** The vitamin C controversy, *Postgrad. Med.,* 67, 64, 1980.
49. **Belfield, W. O. and Stone, I.,** Megascorbic prophylaxis and megascorbic therapy: a new orthomolecular modality in veterinary medicine, *J. Int. Acad. Prev. Med.,* 2, 10, 1975.
50. **Bernik, V.** Vitamin C in high doses and colds, *Rev. Paul. Med.,* 80, 211, 1972.
51. **Berquist, G.,** in *Sven. Laekartidn.,* 37, 1149, 1940.
52. **Bessel-Lorck, C.,** Common cold prophylaxis in young people at ski camp, *Med. Welt.,* 44, 2126, 1959.
53. **Bessey, O. A, Menten, M. L., and King, C. G.,** Pathologic changes in organs of scorbutic guinea pigs, *Proc. Soc. Exp. Biol. Med.,* 31, 455, 1934.
54. **Biilmann, G.,** Ascorbic acid treatment of croupous pneumonia, *Acta Med. Scand. Suppl.,* 123, 102, 1941.
55. **Birkhaug, K. E.,** The role of vitamin C in the pathogenesis of tuberculosis in the guinea pig, *Acta Tuberc. Scand.,* 12, 89, 1938.
56. **Birkhaug, K. E.,** The role of vitamin C in the pathogenesis of tuberculosis, *Acta Tuberc. Scand.,* 13, 45, 1939.
57. **Birkaug, K. E.,** The role of vitamin C in the pathogenesis of tuberculosis in the guinea pig, *Acta Tuberc. Scand.,* 12, 359, 1938.

58. **Blum, K. U.,** Value of ascorbic acid therapy, *Med. Klin.,* 67, 1574, 1972.
59. **Bogen, E., Hawkins, L., and Bennett, E. S.,** Vitamin C treatment of mucous membrane tuberculosis, *Am. Rev. Tuberc.,* 44, 596, 1941.
60. **Bohme H. R., Richter, H., and Fischer, P.,** The effect of drugs on laboratory diagnosis. The effect of ascorbic acid on selected automatic laboratory methods, *Z. Gesamte Inn. Med. Ihre Grenzgeb.,* 34, 574, 1979.
61. **Boissevain, C. H. and Spillane, J. H.,** Effect of synthetic ascorbic acid on the growth of tuberculosis bacillus, *Am. Rev. Tuberc.,* 35, 661, 1937.
62. **Borsalino, G.,** La fragilita capillare nella tubercolosi polmonare e le sue modificazioni per azione della Vitamin C, *G. Clin. Med. (Bologna),* 18, 273, 1937.
63. **Bouhuys, A.,** Colds and antihistaminic effects of vitamin C., *N. Engl. J. Med.,* 290, 633, 1974.
64. **Bourne, G. H.,** Vitamin C and immunity, *Br. J. Nutr.,* 2, 341, 1949.
65. **Bourne, G. H.,** The effect of vitamin C on the healing of wounds, *Proc. Nutr. Soc.,* 4, 204, 1946.
66. **Boyden, S. V. and Andersen, M. E.,** Diet and experimental tuberculosis in the guinea pig, *Acta Pathol. Microbiol. Scand.,* 39, 107, 1956.
67. **Braendon, O. J.,** The common cold: a new approach, *Int. Res. Commun. System,* 7, 12, 1973.
68. **Briggs, M. H. and Briggs, M.,** Vitamin C and colds, *Lancet,* 1, 998, 1973.
69. **Brown, R. G.,** Possible problems of large intakes of ascorbic acid, *JAMA,* 224, 1529, 1973.
70. **Caels, F.,** Contribution to the study of the effect of high doses of vitamin C in oto-rhino-laryngological infections, *Acta Oto Rhino Laryngological Belg.,* 7, 395, 1953.
71. **Calleja, H. B. and Brooks, R. H.,** Acute hepatitis treated with high doses of vitamin C, *Ohio State Med. J.,* 56, 821, 1960.
72. **Cameron, E. and Pauling, L.,** Ascorbic acid and the glycosaminoglycans: an orthomolecular approach to cancer and other diseases, *Oncology,* 27, 181, 1973.
73. **Campanacci, D.,** Current data on the physiopathological and therapeutic importance of vitamin C, *J. Clin. Med.,* 55, 1, 1974.
74. **Carr, A. B., Einstein, R., Lai, L. Y. C., Martin, N. G., and Starmer, G. A.,** Vitamin C and the common cold — using identical twins as controls, *Med. J. Aust.,* 2, 411, 1981.
75. **Carson, M., Corbett, M., Cox, H., and Pollitt, N.,** Vitamin C and the common cold, *Br. Med. J.,* 1, 577, 1974.
76. **Catalano, P. M.,** Vitamin C, *Arch. Dermatol.,* 103, 537, 1971.
77. **Cathcart, R. F.,** Clinical trial of vitamin C, *Med. Tribune,* June 25, 1975.
78. **Chacko, D. D.,** A note on the use of vitamin C in arteriosclerosis, *J. Christ. Med. Assoc. India,* 27, 277, 1952.
79. **Chalmers, T. C.,** Effects of ascorbic acid on the common cold: an evaluation of the evidence, *Am. J. Med.,* 58, 532, 1975.
80. **Chang, T. W. and Snydman, D. R.,** Antiviral agents: action and clinical use, *Drugs,* 18, 354, 1979.
81. **Charleston, S. S., and Clegg, K. M.,** Ascorbic acid and the common cold, *Lancet,* 1, 1401, 1972.
82. **Charpy, J.,** Ascorbic acid in very large doses alone or with vitamin D_2 in tuberculosis, *Bull. Acad. Natl. Med.,* 132, 421, 1948.
83. **Cheraskin, E., Ringsdorf, W. M., Michael, D. W., and Hicks, B. S.,** Daily vitamin C consumption and reported respiratory findings, *Int. J. Vitam. Nutr. Res.,* 43, 42, 1972.
84. **Clegg, K. M.,** Studies associated with ascorbic acid, *Acta Vitaminol. Enzymol.* 28, 101, 1974.
85. **Clegg, K. M. and MacDonald, J. M.,** D-isoascorbic acid investigated in a common cold survey, *Proc. Nutr. Soc.,* 34, 7A, 1975.
86. **Clegg, K. M. and MacDonald, J. M.,** L-ascorbic acid and D-isoascorbic acid in a common cold survey, *Am. J. Clin. Nutr.,* 28, 973, 1975.
87. **Clegg, K. M. and MacDonald, J. M.,** D-isoascorbic acid investigated in a common cold survey. Abstracts of communications, *Br. Nutr. Soc.,* 33, 7A, 1974.
88. **Cochrane, W. A.,** Overnutrition in prenatal and neonatal life: a problem?, *Can. Med. Assoc. J.,* 93, 893, 1965.
89. **Cook, J. D. and Mousen, E. R.,** Vitamin C, the common cold and iron absorption, *Am. J. Clin. Nutr.,* 30, 235, 1977.
90. **Cooper, M. R., McCall, C. E., and DeChatelet, L. R.,** Stimulation of leukocyte hexose monophosphate shunt, activity by ascorbic acid, *Infect. Immun.,* 3, 851, 1971.
91. **Cottingham, E. and Mills, C. A.,** Influence of temperature and vitamin deficiency upon phagocytic functions, *J. Immunol.,* 47, 493, 1943.
92. **Coulehan, J. L.,** Ascorbic acid and the common cold: reviewing the evidence, *Postgrad. Med.,* 66, 153, 1979.
93. **Coulehan, J. L.,** Ascorbic acid and the common cold: reviewing the evidence, *Postgrad. Med.,* 66, 157, 1979.

94. **Coulehan, J. L.**, Ascorbic acid and the common cold: reviewing the evidence, *Postgrad. Med.*, 66, 160, 1979.

95. **Coulehan, J. L., Eberhard, S., Kapner, L., Taylor, P., Rogers, K., and Garry, P.**, Vitamin C and acute illness in Navajo school children, *N. Engl. J. Med.*, 295, 973, 1976.

96. **Coulehan, J. L., Kapner, L., Eberhard, S., Taylor, F. H., and Rogers, K. D.**, Vitamin C and upper respiratory illness in Navajo children: preliminary observations (1974), *Ann. N. Y. Acad. Sci.*, 258, 513, 1975.

97. **Coulehan, J. L., Reisinger, K. S., Rogers, K. D., and Bradley, D. W.**, Vitamin C prophylaxis in a boarding school, *N. Engl. J. Med.*, 290, 6, 1974.

98. **Counsell, J. N. and Hornig, B. H., Eds.**, *Vitamin C (Ascorbic Acid)*, Applied Science, London, 1981.

99. **Cowan, D. W. and Diehl, H. S.**, Antihistaminic agents and ascorbic acid in the early treatment of the common cold, *JAMA*, 143, 421, 1950.

100. **Cowan, D. W., Diehl, H. S., and Baker, A. B.**, Vitamins for the prevention of colds, *JAMA*, 120, 1268, 1942.

101. **Dahlberg, G., Engel, A., and Rydin, H.**, The value of ascorbic acid as a prophylactic against common colds, *Acta Med. Scand.* 119, 540, 1944.

102. **Dainow, I.**, Treatment of herpes zoster with vitamin C, *Dermatologia*, 68, 197, 1943.

103. **Dalton, W. L.**, Massive doses of vitamin C in the treatment of viral diseases, *J. Indiana State Med. Assoc.*, 55, 1151, 1962.

104. **Davies, J. E., Hughes, R. E., Jones, E., Reed, S. E., Craig, J. W., and Tyrrell, D. A.**, Metabolism of ascorbic acid (vitamin C) in subjects infected with common cold viruses, *Biochem. Med.*, 21, 78, 1979.

105. **Debre, R.** L'anergie dans la grippe, *C. R. Seances Soc. Biol (Paris)*, 81, 913, 1918.

106. **DeChatelet, L. R.**, *Ascorbic acid: possible role in phagocytosis*, 62nd Meet. Am. Soc. Biol. Chem., San Francisco, June 18, 1971.

107. **Degeller, J. C.**, Acta brevie neerland physiologie, *Pharmacol., Micribiol.*, 6, 64, 1936.

108. **Demole, V.**, Praktische skorbutkost Nr. 111 aus Haferflocken und Trockenmilch, *Z. Vitaminforsch.*, 3, 89, 1934.

109. **Deni, L.**, Dr. Linus Pauling and vitamin C, *J. Nurs. Care*, 12, 21, 1979.

110. **DeSavitsch, E.**, The influence of organ juice on experimental tuberculosis in guinea pigs, *Natl. Tuberc. Assoc. Trans.*, 30, 130, 1934.

111. **DeWit, J. C.**, Treatment of whopping cough with vitamin C, *Kindergeneeskunde*, 17, 367, 1949.

112. **Diehl, H. S.**, Vitamin C and colds, *New York Times*, December 26, 1970.

113. **Diehl, H. S.**, Vitamin C for colds, *Am. J. Public Health*, 61, 649, 1971.

114. **Drummond, J.**, Recent advances in the treatment of enteric fever, *Clin. Proc. (Cape Town)*, 2, 65, 1943.

115. **Dujardin, J.**, Use of high doses of vitamin C in infections, *Presse Med.*, 55, 72, 1947.

116. **Dugal, L. P.**, Vitamin C in relation to cold, *Ann. N. Y. Acad. Sci.*, 92, 307, 1961.

117. **Dykes, M. H. M. and Meier, P.**, Ascorbic acid and the common cold, *JAMA*, 231, 1073, 1975.

118. **Dymock, I. W., Truck, W. E. G., Brown, P. W., Sircus, W., Small, W. P., and Thomson, C.**, Vitamin C and gastro-duodenal disorders, *Br. Med. J.*, 1, 179, 1968.

119. **Ebner, H.**, On the therapeutic status of influenza and similar infections, *Landartz*, 44, 853, 1968.

120. **Editorial**, Vitamin C and the common cold, *Br. Med. J.*, 1, 606, 1976.

121. **Editorial**, Vitamin C and colds, *S. Afr. Med. J.*, 51, 773, 1977.

122. **Editorial**, Ascorbic acid: immunological effects and hazards, *Lancet*, 1, 308, 1979.

123. **Elliott, B.**, Ascorbic acid: efficacy in the prevention of symptoms of respiratory infection on a Polaris submarine, *Int. Res. Commun. System*, May, 1973.

124. **Elsborg, L.**, Use and misuse of Vitamin C, *Nord. Med.*, 94, 299, 1979.

125. **Elwood, P. C., Hughes, S. J., and St. Leger, A. S.**, A randomized controlled trial of the therapeutic effect of vitamin C in the common cold, *Practitioner*, 218, 133, 1977.

126. **Elwood, P. C., Lee, H. P., St. Leger, A. S., Baird, M., and Howard, A. N.**, A randomized controlled trial of vitamin C, *Br. J. Prev. Soc. Med.*, 30, 193, 1976.

127. **Epstein, L. H. and Masek, B. J.**, Behavioral control of medicine compliance, *J. Appl. Behav. Annal.*, 11, 1, 1978.

128. **Ericsson, Y. and Lundbeck, H.**, Antimicrobial effect, *in vitro* of the ascorbic acid oxidation. I. Effect on bacteria, fungi and viruses in pure cultures. II. Influence of various chemical and physical factors, *Acta Pathol. Microbiol. Scand.* 37, 493, 1955.

129. **Ertel, H.**, Der verlaug der Vitamin C-prophylaxen in Fruhjahr, *Die Ernahrung*, 6, 269, 1941.

130. **Erwin, G. S., Wright, R. and Doherty, G. J.**, Hypovitaminosis C and pulmonary tuberculosis, *Br. Med. J.*, 1, 688, 1940.

131. **Farah, N.**, Enteric fever treated with suprarenal cortex extract and vitamin C intravenously, *Lancet*, 1, 777, 1938.

132. **Faulkner, J. M. and Taylor, F. H. L.**, Vitamin C and infection, *Ann. Int. Med.*, 10, 1867, 1937.

134. **Feldheim, W.,** Grippe-like infections and vitamin C, *Dtsch. Med. Wochenschr.,* 95, 1244, 1970.

135. **Ferreira, D. L.,** Vitamin C in leprosy, *Publ. Med.,* 20, 25, 1950.

136. **Findley, T. W.,** Vitamin C and the common cold, *N. Engl. J. Med.,* 296, 231, 1977.

137. **Fletcher, J. M. and Fletcher, I. C.,** Vitamin C and the common cold, *Br. Med. J.,* 1, 887, 1951.

138. **Floch, H. and Sureau, P.,** Vitamin C therapy in leprosy, *Bull. Soc. Pathol. Exot.,* 45, 443, 1952.

139. **Franz, W. L., Sands, G. W., and Heyl, H. L.,** Blood ascorbic acid level in bioflavonoid and ascorbic acid therapy of common cold, *JAMA,* 162, 1224, 1956.

140. **Gairdner, D.,** Vitamin C in the treatment of whooping cough, *Br. Med. J.,* 2, 742, 1938.

141. **Gander, J. and Neiderberger, W.,** Vitamin C in the treatment of pneumonia, *Muench. Med. Wochenschr.,* 51, 2074, 1936.

142. **Ganguly, R., Khakoo, R., Spencer, J. C., and Waldman, R. H.,** Immunoenhancing agents in prevention and treatment of influenza and other viral respiratory infections, *Dev. Biol. Stand.,* 39, 363, 1977.

143. **Gatti, C. and Gaona, R. J.** Ascorbic acid in the treatment of leprosy, *Arch. Schiffs Trop. Hyg.,* 43, 32, 1939.

144. **Gazaix, M.,** Use of assur in 103 cases of influenza and adenovirus infections, *Sem. Hop.,* 42, 57, 1966.

145. **Gerson, C. D.,** Ascorbic acid deficiency in clinical diseases including regional enteritis, *Ann. N. Y. Acad. Sci.,* 258, 483, 1975.

146. **Getz, H. A., Long, E. R., and Henderson, H. J.,** A study of the relation of nutrition to the development of tuberculosis, *Am. Rev. Tuberc.,* 64, 381, 1951.

147. **Glazebrook, A. J. and Thomson, S.,** The administration of vitamin C in a large institution and its effect on general health and resistance to infection, *J. Hyg.,* 42, 1, 1942.

148. **Goldsmith, G. A.,** Common cold: prevention and treatment with ascorbic acid not effective, *JAMA,* 216, 337, 1971.

149. **Goldstein, M. L.,** High dose ascorbic acid therapy, *JAMA,* 216, 332, 1971.

150. **Gordonoff, T.,** Dorf man wasserlosliche vitamine uberdosieren? Versuche mit Vitamin C, *Schweiz. Med. Wochenschr.,* 90, 726, 1960.

151. **Gorman, J. F.,** The vitamin C controversy, *Postgrad. Med.,* 67, 64, 1980.

152. **Greene, M. R., Steiner, M., and Kramer, B.,** Role of chronic vitamin C deficiency in pathogenesis of tuberculosis in guinea pigs, *Am. Rev. Tuberc.,* 33, 585, 1936.

153. **Greene, M. and Wilson, C. W.,** Effect of aspirin on ascorbic acid metabolism during colds, *Br. J. Clin. Pharmacol.,* 2, 369, 1975.

154. **Greer, E.,** Vitamin C in acute poliomyelitis, *Med. Times,* 83, 1160, 1955.

155. **Gromova, L. I., Chizhikov, D. V., and Minina, S. A.,** Antigrippen tablet technology, *Farmatsiya (Moscow),* 24, 67, 1975.

156. **Grooten, O. and Bezssonoff, N.,** Action of vitamin C on diphtheria toxin, *Ann. Inst. Pasteur,* 56, 413, 1936.

157. **Gsell, O. and Kalt, F.,** Treatment of epidemic poliomyeleitis with high doses of ascorbic acid, *Schweiz. Med. Wochenschr.,* 84, 661, 1954.

158. **Gunzel, W. and Kroehnert, G.,** Experiences in the treatment of pneumonia with vitamin C, *Fortschr. Ther.,* 13, 460, 1937.

159. **Gupta, G. C.D. and Guha, B. C.,** The effect of vitamin C and certain other substances on the growth of microorganisms, *Ann. Biochem. Exp. Med.,* 1, 14, 1941.

160. **Harde, E. and Phillippe, M. M.,** Observations on the antigenic activities of combined diphtheria toxin and vitamin C, *C. R. Acad. Sci.,* 199, 738, 1934.

161. **Harris, A., Robinson, A. B., and Pauling, L.,** Blood plasma L-ascorbic acid concentration for oral L-ascorbic acid dosage up to 12 grams per day, *Int. Res. Commun. System,* December 1973, 19.

162. **Harris, L. J., Passmore, R., Oxon, B. M., Pagel, W., and Berlin, M. D.,** Influence of infection on the vitamin content of the tissues of animals, *Lancet,* 2, 183, 1937.

163. **Hasselbach, F.,** Vitamin C und lungentuberkulose. Veraus-setzungen beabachtungen und erfahrungen bei der behandlung lungentuberkuloser, *Z. Tuberk. Erkr. Thoraxorgane,* 75, 336, 1936.

164. **Hasselbach, F.,** Therapy of tuberculosis pulmonary hemorrhages with vitamin C, *Fortschr. Ther.,* 7, 407, 1935.

165. **Heinrich, H. C.,** Is vitamin C prevention and therapy of virus-caused cold diseases in man justified? *Muench. Med. Wochenschr.,* 110, 1031, 1968.

166. **Heise, F. and Steenken, W., Jr.,** Vitamin C immunity in tuberculosis in guinea pigs, *Am. Rev. Tuberc.,* 39, 794, 1939.

167. **Heise, F., and Martin, G.J.,** Supervitaminosis C in tuberculosis, *Proc. Soc. Exp. Biol. Med.,* 35, 337, 1936.

168. **Heise, F. H. and Martin, G.J.,** Ascorbic acid metabolism in tuberculosis, *Proc. Soc. Exp. Biol. Med.,* 34, 642, 1936.

169. **Hejda, S., Smola, J., and Masek, J.,** Influence of physiological vitamin C allowances on the health status of miners, *Rev. Czech. Med.,* 22, 90, 1976.
170. **Herbert, V. and Jacob, E.,** Destruction of vitamin B12 by ascorbic acid, *JAMA,* 230, 241, 1974.
171. **Hill, R. E. and Ramananda, K.,** "Pink" diarrhoea. Osmotic diarrhoea from a sorbitol-containing vitamin C supplement, *Med. J. Aust.,* 1, 387, 1982.
172. **Hindson, T. C.,** Ascorbic acid for prickly heat, *Lancet,* 1, 1347, 1968.
173. **Hochwald, A.,** Observations on the effect of ascorbic acid on croupous pneumonia, *Wien. Arch. Inn. Med.,* 29, 353, 1936.
174. **Hoffer, A.,** Vitamin C and the common cold, *Can. Med. Assoc. J.,* 105, 901, 1971.
175. **Holden, M. and Molloy, E.,** Further experiments on inactivation of herpesvirus by vitamin C (1-ascorbic acid), *J. Immunol.,* 33, 251, 1937.
176. **Holden, M. and Resnick, R. J.,** In vitro action of synthetic crystalline vitamin C (ascorbic acid) on herpesvirus, *J. Immunol.,* 31, 455, 1936.
177. **Hornig, D.,** Requirements of vitamin C in man, *Trends Pharmacol. Sci.,* 3, 294, 1982.
178. **Horrobin, D. F.,** D. V. T. after vitamin C?, *Lancet,* 2, 317, 1973.
179. **Hughes, R. E.,** Nonscorbutic effects of vitamin C biochemical aspects, *Proc. R. Soc. Med.,* 70, 86, 1977.
180. **Hume, R., Johnstone, J. M. S., and Weyers, E.,** Interaction of ascorbic acid and warfarin, *JAMA,* 219, 1479, 1972.
181. **Hume, R. and Weyers, E.,** Changes in leucocyte ascorbic acid during the common cold, *Scott. Med. J.,* 18, 3, 1973.
182. **Humphrey, G. L., Kemp, G. E., and Wood, E. G.,** A fatal case of rabies in a woman bitten by an insectivorous bat, *Public Health Rep.,* 75, 317, 1960.
183. **Isaksson, B.,** Vitamins 13, vitamin C, *Lakartidningen,* 69, 101, 1972.
184. **Jacob, E., Scott, J., and Brenner, L.,** Apparent low serum vitamin B12 levels in paraplegic veterans, taking ascorbic acid, Proc. 16th Ann. Meet. Am. Soc. Hematol., Chicago, December 1—4, 1973, 125.
185. **Jakowlew, N.,** Zur vitaminnorm, *Ernaehrungsforschung,* 3, 446, 1958.
186. **Jetter, W. W. and Bumbalo, T. S.,** The urinary output of vitamin C in active tuberculosis in children, *Am. J. Med. Sci.,* 195; 362, 1938.
187. **Josewich, A.,** Value of vitamin C therapy in lung tuberculosis, *Med. Bull. Veterans Adm.,* 16, 8, 1939.
188. **Jungeblut, C. W.,** Inactivation of poliomyelitis virus by crystalline vitamin C (ascorbic acid), *J. Exp. Med.,* 62, 517, 1935.
189. **Jungeblut, C. W.,** Inactivation of tetanus toxin by crystalline vitamin C (ascorbic acid), *J. Immunol.,* 33, 203, 1937.
190. **Jungeblut, C. W,** Further observations on vitamin C therapy in experimental poliomyelitis, *J. Exp. Med.,* 65, 127, 1937.
191. **Jungeblut, C. W.,** A further contribution to the vitamin C therapy in experimental poliomyelitis, *J. Exp. Med.,* 70, 327, 1939.
192. **Jungeblut, C. W. and Zwemer, R. L.,** Inactivation of diphtheria toxin in vivo and in vitro by crystalline vitamin C (ascorbic acid), *Proc. Soc. Exp. Biol. Med.,* 32, 1229, 1935.
193. **Kailakov, A. M.,** Concurrence of diabetes mellitus and hypothyrosis in the diencephalic syndrome, *Klin. Med. (Moscow),* 49, 151, 1971.
194. **Kaplan, A. and Zonnis, M. E.,** Vitamin C in pulmonary tuberculosis, *Am. Rev. Tuberc.,* 42, 667, 1940.
195. **Karlowski, T. R., Chalmers, T. C., Frenkel, L. D., Kapikian, A. Z., Lewis, T. L., and Lynch, J. M.,** Ascorbic acid for the common cold: a prophylactic and therapeutic trial, *JAMA,* 231, 1038, 1975.
196. **Keller, H.,** Effective treatment in influenzal infections, *Landarz,* 41, 1489, 1965.
197. **Kent, S.,** Vitamin C therapy: colds, cancer and cardiovascular disease, *Geriatrics,* 33, 91, 1978.
198. **Kent, S.,** Vitamin C therapy: colds, cancer and cardiovascular disease, *Geriatrics,* 33, 99, 1978.
199. **Ketz, H. A.,** Prophylactic effect of vitamin C and vitamin C requirements of a healthy adult, *Dtsch. Gesundheitswes.,* 25, 414, 1970.
200. **Ketz, H. A.,** Vitamin C, prophylactic action, mode of action and requirements, *Nahrung,* 16, 737, 1972.
201. **Kienast, H.,** Treatment of croupous pneumonia with vitamin C, *Muench. Med. Wochenschr.,* 23, 912, 1939.
202. **Kimbarowski, J. A. and Mokrow, N. J.,** Farbige ausfallungsreaktion des Harns nach Kimbarowski als Index der wirkung von ascorbinsaure bei behandlung der virusgrippe, *Dtsch. Gesundheitswes.,* 22, 2413, 1967.
203. **Kinlen, L. and Peto, R.,** Vitamin C and colds, *Lancet,* 1, 944, 1973.
204. **Kirchmair, H.,** Epidemic hepatitis in children and its treatment with high doses of ascorbic acid, *Dtsch. Gesundheitswes.,* 12, 1525, 1957.
205. **Kirchmair, H.,** Treatment of epidemic hepatitis in children with high doses of ascorbic acid, *Med. Monatschr.,* 11, 353, 1957.
206. **Kirchmair, H.,** Ascorbic acid treatment of epidemic hepatitis in children, *Dtsch. Gesundheitswes.,* 12, 773, 1957.

207. **Kleimenhagen, P.**, Effect of ascorbic acid on experimental tuberculosis in guinea pigs, *Z. Vitaminforsch.*, 11, 209, 1941.

208. **Klemola, E.**, Vitamin C and "flu", *Duodecim*, 89, 1273, 1973.

209. **Klenner, F. R.**, The folly in the continued use of a killed polio virus vaccine, *Tri State Med. J.*, February 1959, 1.

210. **Klenner, F. R.**, The vitamin and massage treatment for acute poliomyelitis, *South. Med. Surg.*, 114, 194, 1952.

211. **Klenner, F. R.**, The use of vitamin C as an antibiotic, *J. Appl. Nutr.*, 6, 274, 1953.

212. **Klenner, F. R.**, Observations on the dose and administration of ascorbic acid when employed beyond the range of a vitamin in human pathology, *J. Appl. Nutr.*, 23, 61, 1971.

213. **Klenner, F. R.**, Significance of high daily intake of ascorbic acid in preventive medicine, *J. Int. Acad. Prev. Med.*, 1, 45, 1974.

214. **Klenner, F. R.**, Virus pneumonia and its treatment with vitamin C, *J. South. Med. Surg.*, 110, 36, 1948.

215. **Klenner, F. R.**, The treatment of poliomyelitis and other virus diseases with vitamin C, *J. South. Med. Surg.*, 111, 209, 1949.

216. **Klenner, F. R.**, Massive doses of vitamin C and the viral diseases, *J. South. Med. Surg.*, 113, 101, 1951.

217. **Kligler, I. J. and Bernkopf, H.**, Inactivation of vaccinia virus by ascorbic acid and glutathione, *Nature (London)*, 139, 965, 1937.

218. **Kligler, I. J., Guggenheim, K., and Warburg, F. M.**, Influence of ascorbic acid on the growth and toxin production of Cl tetani and on the detoxication of tetanus toxin, *J. Pathol. Bacteriol.*, 46, 619, 1938.

219. **Kligler, I. J., Leibowitz, L., and Berman, M.**, Effect of ascorbic acid on toxin production of C. diphtheriae in culture media, *J. Pathol. Bacteriol.*, 45, 414, 1937.

220. **Knodell, R. G., Tate, M. A., Akl, B. R., and Wilson, J. W.**, Vitamin C prophylaxis for post transfusion hepatitis: lack of effect in a controlled trial, *Am. J. Clin. Nutr.*, 24, 20, 1981.

221. **Kobza, R., Spioch, F. M., and Kelpacki, J.**, Effect of long term administration of vitamin C on absenteeism in coal miners, *Pol. Tyg. Lek. Wiad. Lek.*, 25, 1007, 1970.

222. **Kodama, T. and Kojima, T.**, Studies of the staphylococcal toxin, toxoid and antitoxin: effect of ascorbic acid on staphylococcal lysins and organisms, *Kitasato Arch. Exp. Med.*, 16, 36, 1939.

223. **Korbsch, R.**, Über die kupierung entzundlich-allergischer zustande durch die L-askorbinasaure, *Med. Klin.*, 34, 1500, 1938.

224. **Kubler, W. and Gehler, J.**, Zür kinetik der enteralen ascorbinsaure-resorption zür berechnung nicht dosisproportionaler resorptionvorgange, *Int. Z. Vitaminforsch.*, 40, 442, 1970.

225. **Kuribayashi, K.**, Effect of vitamin C on bacterial toxins, *Jpn. J. Bacteriol.*, 18, 136, 1963.

226. **Kuttner, A. G.**, Effect of large doses of vitamins A, B, C and D on the incidence of upper respiratory infections in a group of rheumatic children, *J. Clin. Invest.*, 19, 809, 1940.

227. **Lamden, M. P. and Chrystowski, G. A.**, Urinary oxalate excretion by man following ascorbic acid ingestion, *Proc. Soc. Exp. Biol. Med.*, 85, 190, 1954.

228. **Lamy, P. P. and Rotkovitz, I. I.**, The common cold and its management, *J. Am. Pharmacol. Assoc.*, 12, 582, 1972.

229. **Lane, M. M.**, Vitamin C and the common cold, *Nurs. Homes*, 20, 30, 1971.

230. **Langenbusch, W., and Enderling, A.**, Einfluss der vitamine auf das virus der maul-und klevenseuch, *Zentralbl. Bakteriol.*, 140, 112, 1937.

231. **Lechicki, J.**, Has vitamin C any prophylactic effect in colds?, *Wiad. Lek.*, 22, 2234, 1969.

232. **Leibovitz, B. and Siegel, B. V.**, Ascorbic acid and the immune response, *Adv. Exp. Med. Biol.*, 135, 1, 1981.

233. **Lewin, S.**, Vitamin C and the common cold, *Chem. Br.*, 10, 25, 1974.

234. **Lewin, S.**, Evaluation of potential effects of high intake of ascorbic acid, *Comp. Biochem. Physiol. B.*, 47, 681, 1974.

235. **Lewis, T. L., Karlowski, T. R., Kapikian, A. Z., Lynch, J. M., Shaffer, G. W., George, D. A., and Chalmers, T. C.**, A controlled clinical trial of ascorbic acid for the common cold, *Ann. N. Y., Acad. Sci.*, 258, 505, 1975.

236. **Liljefors, I.**, Vitamin C and the common cold, *Lakartidningen*, 69, 3304, 1972.

237. **Loh, H. S., Odumosu, A., and Wilson, C. W. M.**, Factors influencing the metabolic availability of ascorbic acid. I. The effect of sex, *Clin. Pharmacol. Ther.*, 16, 390, 1974.

238. **Lominski, I.**, Inactivation du bacteirophage par l'acide ascorbique, *Soc. Biol.*, 122, 766, 1936.

239. **Lubojacky, M. and Keleti, J.**, Chemoprophylaxis of influenza with Morgalin, *Ther. Hung.*, 17, 127, 1969.

240. **Ludvigsson, J., Hansson, B., and Tibbling, G.**, Vitamin C as a preventative medicine against common colds in children, *S. Afr. Med. J.*, 51, 773, 1977.

241. **McConkey, M. and Smith, D. T.**, The relation of vitamin C deficiency to intestinal tuberculosis in the guinea pig, *J. Exp. Med.*, 58, 503, 1933.

242. **McCormick, W.J.,** Ascorbic acid as a chemotherapeutic agent, *Arch. Pediatr.*, 69, 151, 1952.
243. **McCormick, W. J.,** Vitamin C in the prophylaxis and therapy of infectious diseases, *Arch. Pediatr.*, 68, 1, 1951.
244. **Macon, W. L.,** Citrus bioflavonoids in the treatment of the common cold, *Ind. Med. Surg.*, 25, 525, 1956.
245. **Markwell, N. W.,** Vitamin C in the prevention of colds, *Med. J. Aust.*, 2, 777, 1947.
246. **Martin, G. J. and Heise, F. H.,** Vitamin C nutrition on pulmonary tuberculosis, *Am. J. Dig. Dis. Nutr.*, 4, 368, 1937.
247. **Martin, L. and Lines, J. G.,** The megavitamin scene, *Lancet*, 2, 103, 1974.
248. **Martin, L. F.,** Breakdown of tobacco mosaic virus protein, Proc. 3rd Int. Congr. Microbiol., 1939, 281.
249. **Masek, J., Hruba, F., Neradilova, M., and Hejda, S.,** The role of vitamin C in the treatment of acute infection of the upper respiratory pathways, *Acta Vitaminol. Enzymol.*, 28, 85, 1974.
250. **Masek, J., Neradilova, M., and Hejda, S.,** Vitamin C and resporatory infections, *Cesk. Gastroenterol. Vyz.*, 26, 337, 1972.
251. **Masek, J., Neradilova, M., and Hejda, S.,** Vitamin C and respiratory infections, *Rev. Czech. Med.*, 18, 228, 1972.
252. **Mehnert, H., Forster, H., and Funke, U.,** The effects of ascorbic acid on carbohydrate metabolism, *Ger. Med.* 11, 360, 1966.
253. **Meier, K.,** Vitamin C treatment of pertussis, *Ann. Pediatr. (Paris)*, 164, 50, 1945.
254. **Mick, E. C.,** Brucellosis and its treatment, *Arch. Pediatr.*, 72, 119, 1955.
255. **Miegl, H.,** Acute infections of the upper respiratory tract and their treatment with vitamin C, *Wien. Med. Wochenschr.*, 107, 989, 1957.
256. **Miller, J. Z., Nance, W. E., Norton, J. A., Wolen, R. L., Griffith, R. S., and Rose, R.,** Therapeutic effect of vitamin C: a co-twin control study, *JAMA*, 237, 248, 1977.
257. **Miller, J. Z., Nance, W. E., and Kang, K.,** A co-twin control study of the effects of vitamin C, *Prog. Clin. Biol. Res.*, 24, 151, 1978.
258. **Miller, T. E.,** Killing and lysis of gram-negative bacteria through the synergistic effect of hydrogen peroxide, ascorbic acid and lysozyme, *J. Bacteriol.*, 98, 949, 1969.
259. **Mores, A.** Vitamin C. and common cold, *Cesk. Pediatr.*, 29, 414, 1974.
260. **Mouriquand, G. and Edel, V.** Sur l'hypervitaminose C, *C. R. Soc. Biol.*, 147, 1432, 1953.
261. **Mowle, A. F.,** The vitamin C question. It may be able to stand the test, but can we?, *Aust. Nurs. J.*, 9, 28, 1980.
262. **Murata, A., Kitagawa, K., and Saruno, R.,** Inactivation of bacteriophages by ascorbic acid, *Agric. Biol. Chem.*, 35, 294, 1971.
263. **Murata, A. and Kitagawa, K.,** Mechanism of inactivation of bacteriophage J1 by ascorbic acid, *Agric. Biol. Chem.*, 37, 1145, 1973.
264. **Murata, A.,** Virucidal activity of vitamin C: vitamin C for prevention and treatment of viral diseases, *Proc. 1st Int. Congr. Microbiol. Soc., Sci. Council Jpn.*, 3, 432, 1975.
265. **Myrvik, Q., Weiser, R. S., Houglum, B., and Berger, L. R.,** Studies on the tuberculoinhibitory properties of ascorbic acid derivatives and their possible role in inhibition of tubercle bacilli by urine, *Am. Rev. Tuberc.*, 69, 406, 1954.
266. **Naess, K.,** Vitamin C and the common cold, *Tidsskr. Nor. Laegeforen.*, 91, 666, 1971.
267. **Neuweiler, W.,** Die hypervitaminose und ihre Beziehung zur Schwangerschaft, *Int. Z. Vitaminforsch.*, 22, 392, 1950.
268. **Nitsche, W.,** Vitamin C and its indications, *Med. Monatsschr.*, 27, 454, 1973.
269. **Nordbo, K.,** Vitamin C and the common cold. Report on a recent review of the literature, *Tidsskr. Nor. Laegeforen.*, 96, 894, 1976.
270. **Ormerod, M. J., Byron, M. B., Unkauf, M., and White, F. D.,** Further report on the ascboric acid treatment of whooping cough, *Can. Med. Assoc. J.*, 37, 268, 1937.
271. **Otani, T.,** On the vitamin C therapy of whooping cough, *Klin. Wochenschr.*, 15, 1884, 1936.
272. **Otani, T.** Influence of vitamin C (L-ascorbic acid) upon the whooping cough bacillus and its toxin, *Orient. J. Dis. Infants*, 25, 1, 1939.
273. **Paez de la Torre, J. M.,** Ascorbic acid in measles, *Arch. Argent. Pediatr.*, 24, 225, 1945.
274. **Patterson, J. W.,** The diabetogenic effect of dehydroascorbic and dehydroisoascorbic acids, *J. Biol. Chem.*, 183, 81, 1950.
275. **Pauling, L.,** Ascorbic acid, *Lancet*, 1, 615, 1979.
276. **Pauling, L.,** The case for vitamin C in maintaining health and preventing disease, *Mod. Med. (Chicago)*, July, 68, 1976.
277. **Pauling, L.,** Ascorbic acid and the common cold: evaluation of its efficacy and toxicity, *Med. Tribune*, March 24, 1976.
278. **Pauling, L.,** Vitamin C and colds, *New York Times*, January 17, 1971.
279. **Pauling, L.,** Preventive nutrition, *Med. Midway*, 27, 15, 1973.

280. **Pauling, L.,** Early evidence about vitamin C and the common cold, *J. Orthomolec. Psychiatry.,* 3, 139, 1974.

281. **Pauling, L.,** Are recommended daily allowances for vitamin C adequate? *Proc. Natl. Acad. Sci. U.S.A.,* 71, 4442, 1974.

282. **Pauling, L.,** Vitamin C and common cold, *JAMA,* 216, 332, 1971.

283. **Pauling, L.,** Megavitamin therapy, *JAMA,* 234, 149, 1975.

284. **Pauling, L.,** The significance of the evidence about U.S.A., ascorbic acid and the common cold, *Proc. Natl. Acad. Sci.,* 68, 2678, 1971.

285. **Pauling, L.,** Ascorbic acid and the common cold, *Am. J. Clin. Nutr.,* 24, 1294, 1971.

286. **Pauling, L.,** Evolution and the need for asc. acid, *Proc. Natl. Acad. Sci. U.S.A.,* 67, 1643, 1970.

287. **Pauling, L.,** Ascorbic acid and the common cold, *Scott. Med. J.,* 18, 1, 1973.

288. **Pauling, L.,** *Vitamin C, the Common Cold and the Flu,* Freeman Press, San Francisco, 1976.

289. **Pena, A., Del Arbol, J. L., and Garcia Torres, J. A.,** Effect of vitamin C on excretion of uric acid, *Nutr. Abstr.,* 34, 195, 1964.

290. **Perla, D. and Marmorsten, J.,** Role of vitamin C in resistance, *Arch. Pathol.,* 23, 543, 1937.

291. **Perla, D. and Marmorsten, J.,** Role of vitamin C in resistance, *Arch. Pathol.,* 23, 683, 1937.

292. **Perrault, M. and Dry, J.,** Therapeutics in 1973, *Rev. Prat., Froid,* 23, 5221, 1973.

293. **Petter, C. K.,** Vitamin C and tuberculosis, *Lancet,* 57, 221, 1937.

294. **Pfeiffer, L.,** Ascorbic acid therapy of whooping cough, *Helv. Paediatr. Acta,* 2, 106, 1947.

295. **Pfleger, R. and Scholl, F.,** Diabetes and vitamin C, *Wien. Arch. Med.,* 31, 219, 1937.

296. **Pijoan, M. and Sedlacek, B.,** Ascorbic acid in tuberculous Navajo indians, *Am. Rev. Tuberc.,* 48, 342, 1943.

297. **Pitt, H. A. and Costrini, A. M.,** Vitamin C prophylaxis in Marine recruits, *JAMA,* 241, 908, 1979.

298. **Plate, A.,** Treatment of whooping cough with vitamin C, *Kinderaerztl. Prax.,* 8, 70, 1937.

299. **Poser, E.,** Large ascorbic acid intake, *N. Engl. J. Med.,* 287, 412, 1972.

300. **Preshaw, R. M.,** Vitamin C and the common cold, *Can. Med. Assoc. J.,* 107, 479, 1972.

301. **Querton, M.,** Clinical study of Trimedil in diseases caused by chilling, *Scalpel,* 118, 957, 1965.

302. **Radford, M., DeSavitsch, E., and Sweany, H. C.,** Blood changes following continuous daily administration of vitamin C and orange juice to tuberculosis patients, *Am. Rev. Tuberc.,* 35, 784, 1937.

303. **Rawal, B. D. and McKay, G.,** Inhibition of *Pseudonomas aeruginosa* by ascorbic acid acting singly and in combination with antimicrobials: in-vitro and in-vivo studies, *Med. J. Aust.,* 1, 169, 1974.

304. **Regnier, E.,** The administration of large doses of ascorbic acid in the prevention and treatment of the common cold, Parts I and II., *Rev. Allerg.,* 22, 835, 948, 1968.

305. **Renker, K. and Wegner, S.,** Vitamin C-prophylaxe in der volkswerft stralsund, *Dtsch. Gesundheitswes.,* 9, 702, 1954.

306. **Rhead, W. J. and Schrauzer, G. N.,** Risk of long-term ascorbic acid overdosage, *Nutr. Rev.,* 29, 262, 1971.

307. **Riccitelli, M. L.,** Vitamin C therapy in geriatric practice, *J. Am. Geriatr. Soc.,* 20, 34, 1972.

308. **Rinke, C.,** Vitamin C for bronckospasm, *JAMA,* 245, 548, 1981.

309. **Ritzel, G.,** Critical evaluation of vitamin C as a prophylactic and therapeutic agent in colds, *Helv. Med. Acta,* 28, 63, 1961.

310. **Ritzel, G.,** Ascorbic acid and the common cold, *JAMA,* 235, 1108, 1976.

311. **Robertson, M. G.** Ascorbic acid and the common cold, *JAMA,* 216, 2145, 1971.

312. **Rosenthal, G.,** Interaction of ascorbic acid and warfarin, *JAMA,* 215, 1671, 1971.

313. **Rudra, M. N. and Roy, S. K.,** Haematological study in pulmonary tuberculosis and the effect upon it of large doses of vitamin C, *Tubercle,* 27, 93, 1946.

314. **Ruskin, S. L.,** Calcium cevitamate (calcium ascorbate) in the treatment of acute rhinitis, *Ann. Otol. Rhinol. Laryngol.,* 47, 502, 1938.

315. **Russell W. O. and Calloway, C. P.,** Pathologic changes in the liver and kidneys of guinea pigs deficient in vitamin C, *Arch. Pathol.,* 35, 546, 1943.

316. **Sabin, A. B.,** Vitamin C in relation to experimental poliomyelitis, *J. Exp. Med.,* 69, 507, 1939.

317. **Sabiston, B.H. and Radomski, N. W.,** Health problems and vitamin C in Canadian Northern Military Operations, Defence and Civil Institute of Environmental Medicine Report No. 7, 4-R-1012, Government Printing Office, Ottawa, 1974.

318. **Sadun, E. H., Bradin, J. L., and Faust, E. C.,** Effect of ascorbic acid deficiency on the resistance of guinea pigs to infection with: *Endamoeba histolytica* of human origin, *Am. J. Trop. Med.,* 31, 426, 1951.

319. **Sakakura, Y., Sasaki, Y., Togo, Y., Wagner, H. N., Jr., Hornick, R. B., Schwartz, A. R., and Proctor, D. F.,** Mucociliary function during experimentally induced rhinovirus infection in man, *Ann. Otol. Rhinol. Laryngol.,* 82, 203, 1973.

320. **Samborskaja, E. P.,** Effect of large doses of ascorbic acid on course of pregnancy and progeny in the guinea pig, *Nutr. Abstr. Rev.,* 34, 988, 1964.

321. **Samborskaja, E. P. and Ferman, T. D.**, Mechanism of interruption of pregnancy by ascorbic acid, *Nutr. Abstr. Rev.*, 37, 73, 1967.

322. **Sapozhnikov, I. V., Iakovleva, R. E., Cherezova, L. M., and Ignateva, M. F.**, Nonspecific methods of prophylaxis of influenza and other acute respiratory diseases with dibasole and ascorbic acid, *Vopr. Virusol.*, 4, 42, 1976.

323. **Saracci, R., Bardelli, D., and Mariani, F.**, Clinical trials with vitamin C, *Lancet*, 1, 400, 1975.

324. **Schade, H. A.**, Beitrag zur Frage des einflusses von Vitamin C (L-ascorbinsaure) auf pigmentierungvorgange, *Klin. Wochenschr.*, 14, 60, 1935.

325. **Schneider, G.**, More on the cold war, *Can. Med. Assoc. J.*, 112, 147, 1975.

326. **Schrauzer, G. N. and Rhead, W. J.**, Ascorbic acid abuse: effects of long term ingestion of excessive amounts of blood levels and urinary excretion, *Int. J. Vitam. Nutr. Res.*, 43, 201, 1973.

327. **Schulze, E. and Hecht, V.**, Über die wirkung der ascorbinsaure zur diphtherie-formol-toxoid und tetanus toxin, *Klin. Wochenschr.*, 16, 1460, 1947.

328. **Schwartz, A. R., Togo, Y., Hornick, R., Tominago, S., and Gleckman, R.**, Evaluation of the efficacy of ascorbic acid in prophylaxis of induced rhinovirus 44 infection in man, *J. Infect. Dis.*, 128, 500, 1973.

329. **Schwartz, P. L.**, Ascorbic acid in wound healing — a review, *J. Am. Diet. Assoc.*, 56, 497, 1970.

330. **Schwedt, P. R. and Schwerdt, C. E.**, Effect of ascorbic acid on rhinovirus replication in W1—38 cells, *Proc. Soc. Exp. Biol. Med.*, 148, 1237, 1975.

331. **Sciuk, F.** Boxazin in rheumatic lesions. Therapeutic experiences in a rural practice, *Z. Allg.*, 46, 1086, 1970.

332. **Scott, J. E. and Brenner, L.**, Apparent low serum vitamin B12 levels in paraplegic veterans taking ascorbic acid, Proc. 16th Annu. Meet. Am. Soc. Hematol., Chicago, December 1—4, 1973, 125.

333. **Seifter, J.**, Editorial: ascorbic acid and the common cold, *J. Clin. Pharmacol.*, 15, 509, 1975.

334. **Sessa, T.**, Vitamin therapy of whooping cough, *Riforma Med.*, 56, 38, 1940.

335. **Shadrin, A. S., Araslanova, II., Dekterev, A. N., Gagarinova, V. M., and Tamarkina, K. N.**, Organization and the results of the early ambulatory treatment of influenza patients, *Sov. Med.* 9, 103, 1980.

336. **Shekhtman, G. A.**, On the significance of continuous addition of vitamin C to food in a military sector, *Voen. Med. Zh.*, 3, 46, 1961.

337. **Siegel, B. V.**, Enhanced interferon response to murine leukaemia virus by ascorbic acid, *Infect. Immun.*, 10, 409, 1978.

338. **Sigal, A. and King, C. G.**, The influence of vitamin C deficiency upon the resistance of guinea pigs to diphtheria toxin, *J. Pharmacol.*, 61, 1, 1937.

339. **Sirsi, M.**, Antimicrobial action of vitamin C on *M. tuberculosis* and some other pathogenic organisms, *Indian J. Med. Sci.*, 6, 252, 1952.

340. **Sokolova, V. S.**, Application of vitamin C in treatment of dysentery, *Ter. Arkh.*, 30, 59, 1958.

341. **Spengler, F.**, Vitamin C und der diuretische effekt bei leberzirrhose, *Muench. Med. Wochenschr.*, 84, 779, 1937.

342. **Spero, L. M. and Anderson, T. W.**, Ascorbic acid and common colds, *Br. Med. J.*, 4, 354, 1973.

343. **Stein, W.**, The role of vitamin C and adrenal cortex hormone in the treatment of pneumococcal pneumonias, *Med. Bull. Veterans Adm.*, 18, 156, 1941.

344. **Steinbach, M. M. and Klein, S. J.**, Vitamin C in experimental tuberculosis, *Am. Rev. Tuberc.*, 43, 403, 1941.

345. **Stone, I.**, *The Healing Factor: Vitamin C Against Disease*, Grosset & Dunlap, New York, 1974.

346. **Stone, I.**, The possible role of mega-ascorbate in the endogenous synthesis of interferon, *Med. Hypotheses*, 6, 309, 1980.

347. **Stott, E. J. and Taylor-Robinson, D.**, The common cold, *Practitioner*, 205, 735, 1970.

348. **Sweany, H. C., Clancy, C. L., Radford, M. H., and Hunter V.**, The body economy of vitamin C in health and disease, *JAMA*, 116, 469, 1941.

349. **Sylvestre, J. E. and Giroux, M.**, Vitamin C therapy of pulmonary tuberculosis, *Laval Med.*, 10, 417, 1945.

350. **Szirmai, F.**, Value of vitamin C in treatment of acute infectious diseases, *Dtsch. Arch. Klin. Med.*, 85, 434, 1940.

351. **Taft, G. and Fieldhouse, P.**, Vitamin C and the common cold, *Public Health*, 92, 19, 1978.

352. **Takahashi, Z.**, in *Nagoya J. Med. Sci.*, 12, 50, 1938.

353. **Tanabe, S.**, Vitamin C content of mucous membranes of paranasal sinuses, *Otolaryngology (Tokyo) (Jibi Inkoka)*, 35, 25, 1963.

354. **Taylor, G.**, Vitamins in illness, *Br. Med. J.*, 1, 292, 1973.

355. **Taylor, T. V., Rimmer, S., Day, B., Butcher, J., and Dymock, I. W.**, Ascorbic acid supplementation in the treatment of pressure sores, *Lancet*, 2, 544, 1974.

356. **Tebrock, H. E., Arminio, J. J., and Johnston, J. H.**, Usefulness of bioflavonoids and ascorbic acid in treatment of the common cold, *JAMA*, 162, 1227, 1956.
357. **Thadda, S. and Hoffmeister, W.**, Die bedeutung des C-Vitamins fur infektionsablauf und krankheitsab-wehr, *Z. Klin. Med.*, 132, 379, 1937.
358. **Thomas, H.**, Treatment of fever and influenza infections using Chinavit, *Z. Allg. Med.*, 53, 508, 1977.
359. **Thomas, W. R. and Holt, P. G.**, Vitamin C and immunity: an assessment of the evidence, *Clin. Exp. Immunol.*, 32, 370, 1978.
360. **Tyrrel, D. A. J.**, Cure for all ills?, *Br. Med. J.*, May, 14, 1273, 1977.
361. **Tyrrell, D. A.** The virus causes of coughs and colds, *Helv. Med. Acta*, 47, 52, 1967.
362. **Tyrrell, D. A., Craig, J. W., Meada, T. W., and White, T.**, A trial of ascorbic acid in the treatment of the common cold, *Br. J. Prev. Soc. Med.*, 41, 189, 1977.
363. **Ugarizza, R. G.**, Ascorbic acid in the treatment of leprous specticemia, *Arch. Schiffs Tropen. Hyg.*, 43, 33, 1939.
364. **Vallance, S.**, Relationships between ascorbic acid and serum proteins of the immune system, *Br. Med. J.*, 2, 437, 1977.
365. **Van Alyea, O. E.**, The acute nasal infection, *Nebr. State Med. J.*, 27, 265, 1942.
366. **Vargas-Magne, R.**, Vitamin C in treatment of influenza, *El Dia Med.*, 35, 1714, 1963.
367. **Vermillion, E. L. and Stafford, G. E.**, A preliminary report on the use of cevitamic acid in the treatment of whooping cough, *J. Kan. Med. Soc.*, 39, 469, 1938.
368. **Veselovskaia, T. A.**, Effect of vitamin C on the clinical course of dysentery, *Voen. Med. Zh.*, 3, 32, 1957.
369. **Vilter, R. W.**, Nutritional aspects of ascorbic acid: uses and abuses, *West. J. Med.*, 133, 485, 1980.
370. **Vitorero, J. R. B. and Doyle, J.**, Treatment of intestinal tuberculosis with vitamin C, *Med. Wkly.*, 2, 636, 1938.
371. **Vodopija, I.**, Common cold, *Lijec. Vjesn.*, 92, 1055, 1970.
372. **Walker, E.**, Megavitamin research data incomplete but promising, *Mod. Nurs. Home*, 30, 4, 1973.
373. **Walker, G. H., Bynoe, M. L., and Tyrrell, D. A. J.**, Trial of ascorbic acid in prevention of colds, *Br. Med. J.*, 1, 603, 1967.
374. **Wasilewski, A.** Influenza therapy in the practice, *Z. Allg. Med.*, 47, 62, 1971.
375. **Widenbauer, F.**, Toxische nebenwirkungen von ascorbinsaure C-hypervitaminose? *Klin. Wochenschr.*, 15, 1158, 1936.
376. **Williams, H. E.**, Calcium nephrolithiasis and cellulose phosphate, *N. Engl. J. Med.*, 290, 224, 1974.
377. **Willis, G. C.** The influence of ascorbic acid upon the liver, *Can. Med. Assoc. J.*, 76, 1044, 1957.
378. **Wilson, C. W.**, Colds and vitamin C., *Ir. Med. J.*, 68, 511, 1975.
379. **Wilson, C. W.**, Ascorbic acid and colds, *Br. Med. J.*, 2, 698, 1967.
380. **Wilson, C. W.**, Clinical pharmacological aspects of ascorbic acid, *Ann. N.Y. Acad. Sci.*, 258, 355, 1975.
381. **Wilson, C. W.**, Vitamin C and fertility, *Lancet*, 2, 859, 1973.
382. **Wilson, C. W.**, The role of vitamin C and in the prophylaxis of common cold and hay fever, *Acta Vitaminol. Enzymol.*, 2, 120, 1980.
383. **Wilson, C. W.**, Vitamin C for common colds, *Lancet*, 1, 586, 1976.
384. **Wilson, C. W.** Vitamin C and the common cold, *Br. Med. J.*, 1, 1470, 1976.
385. **Wilson, C. W.**, Ascorbic acid and the common cold, *Practitioner*, 215, 343, 1975.
386. **Wilson, C. W.**, Ascorbic acid function and metabolism during cold, *Ann. N.Y. Acad. Sci.*, 258, 529, 1975.
387. **Wilson, C. W.**, Colds, ascorbic acid metabolism and vitamin C, *J. Clin. Pharmacol.*, 15, 570, 1975.
388. **Wilson, C. W.**, Vitamin C tissue saturation metabolism and desaturation, *Practitioner*, 212, 481, 1974.
389. **Wilson, C. W.**, Vitamin C and the common cold, *Br. Med. J.*, 1, 669, 1971.
390. **Wilson, C. W.**, The common cold and Vitamin C prophylactic, therapeutic, metabolic and functional aspects, *Acta Vitaminol. Enzymol.*, 28, 96, 1974.
391. **Wilson, C. W. and Greene, M.**, The transfer and utilization of vitamin C in human tissues, *Proc. Nutr. Soc.*, 33, 109A, 1974.
392. **Wilson, C. W. and Greene, M.**, The relationship of aspirin to ascorbic acid metabolism during the common cold, *J. Clin. Pharmacol.*, 18, 21, 1978.
393. **Wilson, C.W., Greene, M., and Loh, H. S.** The metabolism of supplementary vitamin C during the common cold, *J. Clin. Pharmacol.*, 16, 19, 1976.
394. **Wilson, C. W. and Loh, H. S.**, Vitamin C metabolism and the common cold, *Eur. J. Clin. Pharmacol.*, 7, 421, 1974.
395. **Wilson, C. W. and Loh, H. S.**, Ascorbic acid and upper respiratory inflammation, *Acta Allergol.*, 24, 367, 1970.
396. **Wilson, C. W. and Loh, H. S.**, Vitamin C and colds, *Lancet*, 1, 1058, 1973.
397. **Wilson, C. W. and Loh, H. S.**, Ascorbic acid and the common cold, *Br. Med. J.*, 4, 166, 1973.
398. **Wilson, C. W. and Loh, H. S.**, Common cold and vitamin C, *Lancet*, 1, 638, 1973.

399. **Wilson, C. W., Loh, H. S., and Foster, F. G.**, Common cold symptomatology and vitamin C, *Eur. J. Clin. Pharmacol.*, 6, 196, 1976.
400. **Wilson, C. W., Loh, H. S., and Foster, F. G.**, The beneficial effect of vitamin C on the common cold, *Eur. J. Clin. Pharmacol.*, 6, 26, 1973.
401. **Woolfe, S. N., Hume, W. R., and Kenney, E. B.**, Ascorbic acid and periodontal disease: a review of the literature, *Periodontal Abstr.*, 28, 44, 1980.
402. **Woolstone, A. S.**, Treatment of the common cold, *Br. Med. J.*, 2, 1290, 1954.
403. **Yakovlev, N. N.**, Vitamin B and C requirements during participation in sports, *Vopr. Pitan.*, 8, 3, 1958.
404. **Yonemoto, R. H., Chretien, P. B., and Fehniger, T. F.**, Enhanced lymphocyte blastogenesis by oral ascorbic acid, *Am. Soc. Clin. Oncol.*, 288, 1976.
405. **Zlydnikov, D. M.**, Experience with clinical trials of the new anti-influenza drug rimantadine, *Vrach.Delo.*, 4, 15, 1981.
406. **Zlydnikov, D. M. and Chepuk, E. B.**, Clinical course and principles of intensive treatment of hypertoxic forms of influenza complicated by pneumonia, *Ter. Arkh.*, 48, 45, 1976.
407. **Zlydnikov, D. M., Romanov, I., Rumel, N. B., Vasilevskaia, N. M., and Geiker, V.I.**, Therapeutic effectiveness of rimantadine and antigrippen against A1(H1N1) influenza in the 1977—1978 epidemic, *Vrach. Delo.*, 6, 109, 1981.
408. **Zureick, M.**, Treatment of shingles and herpes with vitamin C intravenously, *J. Prat.*, 64, 586, 1950.
409. **Abbott, P., Abrahams, M., Adams, M. S. M.**, et al., Ineffectiveness of vitamin C in treating coryza, *Practitioner*, 200, 442, 1968.
410. **Albanese, A. A., Wein, E. H., and Mata, L. A.**, Improved method for determination of leucocyte and plasma ascorbic acid of man with application to studies of nutritional needs and effects of cigarette smoking, *Nutr. Rep. Int.*, 12, 271, 1975.
411. **Briggs, M.**, Blood platelet biochemistry in women receiving steroid contraceptives, *Haematologica*, 7, 347, 1973.
412. **Costello, J., Hatch, M., and Bourke, E.**, Enzymatic method for the spectrophotometric determination of oxalic acid, *J. Lab. Clin. Med.*, 87, 903, 1976.
413. **Roe, J. H. and Kuether, C. A.**, Determination of ascorbic acid in whole blood and urine through 2,4-dinitro phenyl hydrazine derivative of dehydro ascorbic acid, *J. Biol. Chem.*, 147, 399, 1943.

Chapter 4

VITAMIN SUPPLEMENTS AND INTRAUTERINE GROWTH

Karl Kristoffersen and John Rolschau

TABLE OF CONTENTS

I. INTRODUCTION

This chapter provides a survey of the recent studies dealing with the influence of vitamins on the reproductive process expressed by intrauterine development and growth of the fetus. The vitamins available to the fetus comprise both natural vitamin content in food, fortification of food, and supplements taken as special pharmaceutically prepared drugs.

Intrauterine growth can be evaluated quantitatively by the rate of fetal weight and length acceleration, and qualitatively by the development of organs during the embryogenesis. This can be measured by the incidence of spontaneous abortions and embryopathies or malformations.

To measure the influence of vitamins on intrauterine growth is a comprehensive task because so many factors are implicated in the growth. At first we will enumerate the known possible influencing factors, procedures, and pitfalls which must be taken into account before attempts to reach conclusions can be made. Folic acid deficiency is the most common vitamin deficiency in pregnancy and the vitamin being paid most attention as the possible cause of impaired fetal development. Therefore, we will describe investigations dealing with folic acid deficiency and only very briefly mention other vitamins.

II. FOLIC ACID IN FOOD

The content of folic acid in food variates greatly with geography, meteorology, and with the way the food is prepared.

A. Compounds, Contents, and Analyses

Folic acid, folacin, or folate is the general term for the constituents in folic acid biological activity. However, folates in food comprise both pteroyl glutamic acid which is a mono-glutamate, and formyl- and methyl-polyglutamate with up to ten glutamate residues. The polyglutamates have to be decomposed by enzymes to mono-, di-, or triglutamates before absorption and subsequent utilization can take place. The results of measurement of the total amount of biological usable folates in food therefore depend on the method employed. Several microbiological methods have been described. *Lactobacillus casei* is most often used, but the pretreatment before stimulation of the bacteria varies from heating to 121°C in 15 min.,[1] autoclaving,[2] to heating in boiling water for 10 min.[3] The same problems are faced in measuring folates in red blood cells, serum, or plasma. Avery and Ledger[4] state that heparinized blood is hemolyzed in hypotonic sodium ascorbate solution and incubated at 37° C for 24 hr to release conjugated compounds.

Chanarin et al.[5] measured red cell folate by the *L. casei* method and the blood was diluted tenfold in a 0.1 *M* phosphate ascorbate buffer of pH 6 (Hansen, 1964). They found that this method gave red cell folate results one third lower than that obtained when blood was diluted in an aqueous solution of ascorbate. These are examples possibly explaining some of the different results in the literature. The term "free folic acid" is used concerning the folate which can be utilized directly by *L. casei,* while total folate also comprises compounds that have to be released by enzymatic activity before microbiological utilization is possible.

Analyses of folate content in food reveal that free folic acid content ranges from 53 to 296 μg/day (mean 160), while total folate including conjugated forms ranges from 197 to 1615 μg/day (mean 676).[1] Others have found mean values on 206 μg of free folate and 242 μg of total folates.[6] Many estimations of vitamin content in a diet have been performed by using food tables. In Denmark such an analysis over 1 year showed a mean folate intake of 70 μg/day by teenagers and of 82 μg/day by pregnant women.[7] This is only 10% of the recommended daily allowance for pregnancy.[8]

Lowenstein[9] judged from food composition tables that daily intake in the U.S. was 82 to 92 μg in uncooked food. Cooking reduced this content to 10 to 65% or even more.[10]

The free folate content of Iranian bread, *Bazari and Sangak,* which is leavened flat bread, was found to be from 0.34 to 0.71 µg/g as compared with only 0.13 µg/g in white bread, and 0.09 µg/g in oatmeal.[11] The large content of fibers in bread is found not to influence the absorption rate of folate.[11]

The utilization of folate also depends on the composition of the food. Folic acid is better absorbed when cooked in maize meal than from bread, and 300 µg free folate given in a tablet correspond to 900 µg given in bread.[12] Andersson et al.[13] demonstrated that 85% of free folate was absorbed by women who were not pregnant, but much less seemed to be absorbed from the conjugated forms of folate. Exact information, however, is not available. Dietary histories are useless in the underdeveloped countries because people sometimes eat from a communal pot.[14]

Other investigators[15] have used the *Streptococcus faecalis* method for measuring folic acid, and results obtained by this method are much lower than those obtained by the *Lactobacillus casei* method.[1,2] This is because *L. casei* can utilize both mono-, di- and triglutamates, while *S. faecalis* demands monoglutamate for stimulation. The ratios between the results found by the two methods, however, is widely different, 1/3 and 1/14, respectively.[1,2]

B. Seasonal Variations

Seasonal variations are found in serum folate concentrations,[14,16] and in red cell folate.[17] Seasonal variations in the incidence of megaloblastic anemia probably reflect seasonal variations in folate intake.[5,18]

III. HUMAN FACTORS INFLUENCING UTILIZATION OF FOLIC ACID

A. Absorption, Transport, and Consumption

In consequence of the mentioned factors, the cooking, eating, and drinking habits are of great importance for a sufficient supply of folic acid. Moreover, the enzymes in food, the enzymatic activity, and microbiologic flora in the GI tract may influence the amount of absorbed folates.

The absorption takes place in the jejunum and an enzyme, pteroyl polyglutamate hydroxylase, decomposes the polyglutamates for optimal utilization of folates.[19] This process, however, is generally of minor importance to the supply of folate.

An enzymatic transformation to biologically active tetrahydrofolic acid by dihydrofolate reductase takes place in the liver, which is the most important depot for folic acid. In the transport to the tissues folic acid is bound reversible to albumin and less reversible to specific binders.[20,21] The increased level of folic acid is seen in urinary infections and is perhaps due to increased consumption.[22]

In some diseases of malabsorption low folate concentrations are found.[23] Great alcohol consumption is linked to a low folate state either because of low intake or by its effect on metabolic function.[24]

B. Drugs' Interaction with Folic Acid

Many drugs are interfering with the folate state in different ways. Acetyl salicylic acid is in vitro found to decrease the folic acid binding to proteins.[25] Salazosulfapyridine inhibits absorption and antiepileptics probably act by competitive interaction in folic acid metabolism and/or by increasing the requirement of folic acid.[23,26,27]

Another group of antifolate drugs are inhibitors of the dihydrofolic reductase, which transforms the folic acid to tetrahydrofolic acid, the form necessary for the ribonucleotide synthesis in the cell. The greatest affinity to reductase is methotrexate which is used for cancer chemotherapy and very often causes severe folic acid deficiency.[28] Trimetoprim with lower affinity to human reductase is used together with sulfa in treatment of infections and may have an adverse effect on the fetus in the beginning of pregnancy.

C. Congenital Errors of Folic Acid Metabolism

About ten different forms of inborn errors of folic acid metabolism are found.[29] The mechanisms acting in these errors are different deficiencies of enzymes or carrier proteins, so that the synthesis of nucleotides and certain amino acids are compromised. Best known is the formimino transferase deficiency syndrome,[30] causing mental retardation, opticus atrophy, and EEG abnormalities.

IV. CONDITIONS IN PREGNANCY ALTERING VITAMIN STATUS, ESPECIALLY IN FOLATES

A. Biochemical Changes in the Mother During Pregnancy

For understanding the differences found in the folate state between pregnant and nonpregnant women it is necessary to know the considerable changes during pregnancy in hemodynamics, renal and tissue clearance, the kinetics of carrier proteins for the role of placenta in the active maternofetal transport and in the metabolism of vitamins, and the kinetics of folic reductase in the cells in the synthesis of DNA and RNA.

Decreasing concentrations of both serum and erythrocyte folic acid during pregnancy in women not supplemented with folic acid are almost invariably found.[1,31-34] Explanations for these findings may be

1. Hemodilution during pregnancy.
2. Lower intake of folic acid in the first trimester because of nausea.
3. Impaired absorption during pregnancy.[15]
4. Increased requirements from the rapid growth of the fetus, and the hyperplasia of various maternal tissues.
5. Serum folic acid clearance increases during pregnancy, reflecting the increased requirement.[15] In twin pregnancies there are even higher clearances than in single pregnancies, and it is extremely high in patients with megaloblastic anemia.[15]
6. By measuring folic acid in the blood of pregnant women, some underestimation may be found because of displacements in the biochemical composition of blood.[35] Arekull et al.[36] found increasing values of folate-binding protein during pregnancy and decreasing values of folic acid in serum. There is significant negative correlations between corresponding values of these two components. They claim that the mechanism in these kinetics is not exactly known, but propose a hormonal effect. Fleming[37] found that the urinary excretion of folate was higher in pregnant than in nonpregnant women. He postulated that this could be a factor contributing to the lowering of folate activity during pregnancy. This explanation was followed up by Arekull et al.[36] They postulated that the greater demand of folate causes a stimulation of synthesis of folic acid binding protein (FABP). It would have been interesting to perform recovery experiments, where increasing amounts of FABP were mixed to folic acid and to find out if there was an effect on the results.
7. The very great increase in hormone production during pregnancy, especially in estrogens, stimulates the production of special globulins, which act as carriers for hormones and vitamins important for the fetus.

Chromatographic determinations of globulins in nonpregnant women showed that 26% of folic acid was protein bound. In pregnant women the corresponding percentages had increased from 92 to 100%. Most of folic acid was bound to transferrin and α-2-macroglobulin. A few weeks after delivery, a heavy fall in the percentage of protein-bound folic acid takes place.[35]

B. Transplacental Maternofetal Transport of Folic Acid

Plasma concentrations of folic acid in umbilical cord blood are found to be 4 to 5 times that of maternal serum folic acid from the twenty-second to the forty-second week of pregnancy.[38] Red cell folic acid was about two times that in their mothers' blood from the twenty-second to the thirty-fourth week; during the last weeks of pregnancy this ratio increased to 3:1.[38] In other investigations the ratio between cord blood and maternal plasma was 3.5:1 (9.8 contra 2.8 ng/mℓ) and 2:1.[39,40]

Kamen and Caston[41] found by chromatography a folate binder in human umbilical cord serum containing both free and complex folate, the latter being bound to a binder with a very high affinity, in concentrations corresponding to 60 ng of 5-methyl-tetrahydrofolic acid per milliliter of serum. This is consistent with the finding that the placenta is able to carry folic acid from mother to child against a very high concentration gradient. Kaminetzky[42] corroborated this finding and suggests that a modification of vitamins takes place in the placenta. It would be of great interest to compare the two binders from maternal and cord blood concerning the ability to bind identical folic acid components to elucidate the nature of this very important "trap" effect of placenta.

V. INTRAUTERINE FETAL GROWTH

A. The Evaluation of Normal Intrauterine Growth

Normal growth is usually estimated with reference to a cross-section material made of average skeletal measures as weight, length, head circumference, and biparietal diameter. The latter can also be estimated on longitudinal curves measured throughout pregnancy, and the form of the curve has to be parallel to that of the reference curve. Those reference curves, cross-sectional or longitudinal, are constructed from consecutive or selected materials and naturally include known ethnic differences.[43]

There are other known factors which affect fetal growth and which can be clarified when constructing the curves. These are the mother's prepregnant weight, height, parity, and the infant's sex. Furthermore, there are known factors such as tobacco consumption, use of alcohol and narcotics in significant doses, stress, and hard physical work, which with a high degree of probability will retard intrauterine growth of the fetus.[43] It is often difficult to see how much these factors affect the growth curves.

The curves are made in definite geographic areas and include the normal differences in intake of calories and vitamins. These are further dependent upon socioeconomic and seasonal variations. If the value of these factors is clear in the used reference curves, an estimate of an individual growth can be made. The approximate value of the effect on infant weight by sex, parity, and tobacco consumption are made of concurrent authors.[45,46] The dependent geographical factors are difficult to expose since they infiltrate the whole sample and often are only revealed in intervention trials with one of the variables. The significance of the single factors differs from one population to another and reveals the complexity of the causes of interuterine growth retardation.

B. Parameters Used for Evaluation of Intrauterine Growth

During pregnancy growth can be roughly estimated by indirect means. Widely used are the mother's weight increase, uterine growth judged from abdominal circumference, or symphysis-fundus measure. The placental weight can to some extent be estimated by measuring a hormone in it (human placental lactogen), and it correlates roughly with fetal weight.

The most valuable parameter for estimating fetal size during pregnancy is ultrasound measurements of fetal dimensions.

Postnatally the accepted parameters are weight, length, head circumference, and biparietal diameter. Included in addition to these skeletal dimensions are measurements of subcutaneous fat as an expression of the offered amounts of calories. Organ size, expressed as cell number by measurements of DNA[47] seems more relevant as it is closer correlated with the final outcome: the physiologically and psychologically normal child.

C. The Pathogenesis of Intrauterine Growth Retardation

As mentioned earlier, it is difficult to judge whether the growth of a fetus is normal or better optimal. The greatest possible size is hardly optimal. A well-known example is that women who are overweight as well as diabetic women deliver heavy children with a thick subcutaneous fat layer, presumably as an expression of an increased concentration of glucose.

The optimal intrauterine growth is a growth resulting in children who throughout childhood have normal height and weight and are free of malformations and with normal intelligence and behavior. The tentative pathogenesis is that a noxe, or a lack of nutrients leads to impairment of fetal growth, and fewer or smaller cells in the vital organs developing during childhood result in subnormal physical and mental performance. Known factors in the mother which impair the growth of the child are toxemia, hypertension, nephropathia, heart disease, immunizations, and other serious chronic diseases. Furthermore, abuse of tobacco, narcotics, and alcohol also impair growth. Various degrees of growth retardations can result from calorie, protein, and vitamin deficiencies.

A central point in the mechanism of growth retardation is the decreased blood supply to the uterus,[48] the possible subsequent decreased supply of oxygen and nutrients, and a decreased exchange of metabolites.

The placenta seems to play an independent role in some forms of growth retardation. For example, it is hypertrophied in low oxygen pressure.[49] Under other conditions it is impaired simultaneously with the fetus. Some placental abnormalities such as inflammation[50] and circumvallation[51] seem to be a pathogenetic link. That placenta plays a solitary role in fetal nutrition is supported by the fact that vitamins are found in higher concentrations in fetal than in maternal blood.[42] The mechanisms in this selective transport, with a possible correlation to pathoanatomic findings, are unknown.

The concentration gradient through the placental membrane of products in the intermediate metabolism is a main issue for growth and growth retardation.

D. The Significance of Intrauterine Growth Retardation

When prophylactic and therapeutic treatments of comprehensive character for the pregnant woman (i.e., bedrest for several months) are used to prevent pronounced growth retardations, it is desirable to be certain of the single steps in the pathogenesis. We know that there are several types of growth retardation,[45,52] but it is unclear which of them lead to sequelae for the infant.

During the famine in Holland[53] a calorie reduction in the whole population led to a decreased birth weight of around 400 g. After 19 years they had no detectable sequelae. Douglas and Gear (1976)[54] found no damage in low birth weight infants after a 15 year study and similarly Fitzhardinge and Steven (1972)[55] found that full-term, small-for-date infants had normal growth after 6 years. Other investigators[56] found slower growth at 4 years in a similar group of infants. Examinations are more conclusive in animal experiments[57] where guinea pigs were undernourished in utero and well nourished after birth. About 20% of the guinea pigs showed a definite reduction in the number of brain cells when they reached maturity.

In conclusion, the consequences of intrauterine growth retardation later in life are uncertain. It is possible that some of the etiological groups of undergrown infants have sequelae, but the groups are not identified.

VI. THE ASSESSMENT OF FOLIC ACID DEFICIENCY AND REQUIREMENT IN PREGNANCY

A. Megaloblastic Anemia

The great quantitative differences in folate state in pregnancy are geographically determined and are found to be good in Australia,[58,59] North America,[4] Norway,[33] and partially in England,[5,15] but poor in South Africa in the Bantus,[34] and India.[16,60] Clinically serious folate deficiency is found only in countries with insufficient folate intake caused by a poor diet and is characterized by a high incidence of megaloblastic anemia. The criteria used for this diagnosis, however, vary from center to center to such a degree that it is impossible to compare incidences.[61] The incidences vary from a few per thousand to 50%, but even in countries with folate-sufficient food there are a few cases of megaloblastic anemia. The concentration of folate in the blood from patients with megaloblastic anemia is very low, and the anemia usually becomes evident in the third trimester of pregnancy or in the puerperium, where the lowest values of folate are found: serum concentrations of labile folic acid from 0.1 to 3.4 ng/mℓ.[62] The measurement of both serum and erythrocyte folates, however, are not ideal for discrimination between normal and megaloblastically anemic mothers. Many pregnant women with normal erythropoiesis and no anemia also have low values of serum folate (under 3 ng/mℓ).

Fortification of food with folates will reduce the incidence of megaloblastic anemia considerably, but even with high doses, e.g., 15 mg/day, occasionally patients with megaloblastic anemia may be seen. Therefore, we propose the hypothesis that there are other pathogenetic steps in folate deficiency. Perhaps a congenital enzyme defect which only gives symptoms at relatively low folic acid concentrations might be the cause.

With the very wide range of serum and erythrocyte folate concentrations in normal pregnancies it is quite impossible to make statements or conclusions as to the definition of folate deficiency from the known methods of measuring folate. By giving supplements of folate it is possible to double the depots of folate in red cells as compared to the "normal" or more correctly expressed "common" values met with in the population. The optimal values are not known and we will discuss this problem in relation to the co-variation and intervention trials.

B. Birth Weight and Folate State

1. Co-Variance Between Birth Weight and Folate State

A most severe deficiency of folate invariably causes growth retardation and death of the fetus and/or malformations of different kinds, as described later. Thus, there may be a correlation between cellular growth and folate state. A positive correlation between serum folate concentrations in mothers at full term and the birth weight of their infants has been found.[22]

Hibbard[63] found a significant positive correlation between red cell folate in early pregnancy and birth weight; the same has been found at full term by others.[17,64-66] Moreover, Hibbard[63] also found a positive correlation to placental weight; other investigators have not found any such relationship.[5,59] The correlation between birth weight and erythrocyte folate is affected by the varying values of erythrocyte folate throughout pregnancy.

Ek (1982)[66] found a significant decrease of about 15% ($p < 0.002$) in erythrocyte folate in pregnant women from the thirtieth week to full term. These women had no folic acid supplements during pregnancy. In spite of this he found the red cell concentrations of folate

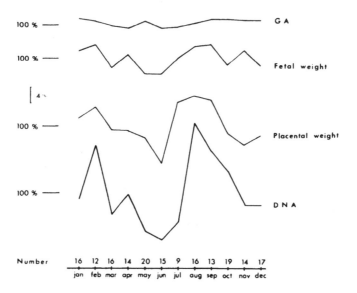

FIGURE 1.   The average values for gestational age, fetal weight, placental
weight and content of DNA over 12 months. The curves are drawn in such a
manner that the fluctuations, measured in millimeters, comprise the same
percentage of the mean values for all four curves.

in mothers to infants with a birth weight of $\leqslant$ 3000 g was 320.0 $\pm$ 26.8 nmol/ℓ, and the
concentration in mothers to infants with a birth weight of from 3000 to 4000 g was 330.7
$\pm$ 16.7 nmol/ℓ; this is a significant difference ($p < 0.05$). Gestational ages were from 38
to 41 completed weeks.

Rolschau et al. (1979)[17] found a positive correlation between erythrocyte folate concentrations and birth weight in 35 women with normal pregnancy and delivery of a single child
with a gestational age of 281 $\pm$ 12 days. In these studies folate state was found to be normal
as evaluated from the hematologic parameters.

2. Intervention Trials with Folate Supplement

Folic acid supplementation during pregnancy has been said to increase the birth weights
of the infants.[17,60,67] Other investigators[58,59,68,69] have not found any effect on birth weight
in such supplementation trials. In prospective folate supplement studies, it is of paramount
interest to consider all possible variables influencing birth weight, including seasonal variations.

In a study from Denmark[70] which comprised 181 delivering women randomly selected
from the population and with identical gestational age through the year, a seasonal variation
was found both in the birth weight, placental weight, and the DNA content of the placenta
(Figure 1). The lowest values were found in May—June and the highest in August—
September. In an intervention trial,[17] we consecutively selected pregnant women attending
the antenatal clinic at Odense University Hospital in the month of January, when all were
in the twenty-first to the twenty-fifth gestational week. They were matched two and two
according to parity, tobacco consumption, prepregnant weight, housing conditions, and age,
and were allotted to two groups. The first group was given 250 mg of ferrofurate + 5 mg
folic acid and the second group was given similar tablets without folic acid. The differences
in birth weight, placental weight, and DNA in the two groups are seen in Table 1. In all
three parameters the highest values were found in the supplemented group. The differences
in birth weight between the two groups was 407 g or 12.7%, which was significant ($p <$
0.01). The conclusion was that a supplement of 5 mg folic acid daily during the final 15 to
20 weeks of gestation to pregnant women who reach full term in the first part of June,

Table 1
DIFFERENCES IN FETAL AND PLACENTAL WEIGHT AND PLACENTAL DNA IN THE FOLIC ACID-TREATED AND THE NONTREATED GROUP[17]

		Mean	T-value	Difference (%)	Significance
Fetal weight (g)	+ Folic acid	3610	2.958	12.7	$p < 0.01$
	− Folic acid	3203			
Placental weight (g)	+ Folic acid	459	1.753	11.9	N.S.
	− Folic acid	410			
Placental DNA (mg)	+ Folic acid	712	1.398	11.1	N.S.
	− Folic acid	641			

Note: N.S. = not significant.

increased birth weight significantly. A similar result was found by Baumslag et al. (1970)[67] and Iyengar and Rajalakshmi (1975).[60] In the first trial, the supplement was 5 mg of folic acid daily. In the next trial, 0.5 mg was given daily. Both trials were performed in populations with very poor folate states. In other studies not showing such relationships, three trials were performed in Australia.[58,59,68]

In one of these trials,[58] 692 consecutively elected pregnant women were randomized into two groups: one group was given a supplement of 5 mg folic acid and the other one placebo. In the other trial,[59] 146 pregnant women were randomized into 5 groups and given a supplement: group I— placebo, group II — iron, group III — 0.5 mg folic acid, group IV — iron + 0.5 mg folic acid, and group V — iron + 5 mg folic acid. No significant differences were found in any of these well-designed trials. One could criticize the lack of mention of tobacco consumption. No control of folate state was done in one trial,[58] and compliance was subsequently untested. In the other trial,[59] the red cell folate state prior to supplementation was better than that obtained in the Danish material.

Fletcher et al.[69] performed a double-blind trial of 643 pregnant women in London. They were randomized into two groups, in which one received a placebo and the other was given 5 mg folic acid daily from the fourteenth week. No difference in gestational age, birth weight, and placental weight was found. Red cell folate was not measured, but serum folate was a little higher in the nonfolate supplemented group before the start of the trial (7.6 vs. 6.3 ng/mℓ).

3. Discussion

It seems controversial that some investigators find a significant increase in birth weight by supplement of folic acid, and others do not. However, there may be a rational explanation for these discrepancies. It is probable that the folate state is better in Australia than in Denmark. The red cell folic acid before the trials[17,59] was found to be 75% higher in the Australian than in the Danish material. In both trials a supplement of 5 mg folic acid per day was given from about the twentieth week until term. In Australia the increase of red cell folate was 3.5 to 5.8 times and in Denmark only 1.5 to 1.7 times. Moreover, the women in Australia received 50 mg ascorbic acid (AA) per day from their initial attendance, which might have increased the absorption rate of folic acid. The birth weight and fetal well being can be compromised by an unfavorable mutual relationship between the compounds of food: carbohydrate, protein, fat, and minerals.

In a large, randomized, controlled trial of nutritional supplements,[53] 1051 pregnant women were recruited to the study according to certain criteria, and allocated into three groups. All three groups were given staple food free and were also given some vitamin supplement.

Among these were 350 μg of folic acid. Group I was given a supplement of 40 g of casein, 55 g of carbohydrates, 8.6 g of fat, and an abundance of minerals. Group II received a supplement of 6 g of casein, 57.4 g of carbohydrates, 7.6 g of fat, and smaller amounts of minerals. Group III received no further supplementation. The gestational age, the perinatal mortality, and birth weight were registered.

The only significant favorable effect of high protein intake was found in group I compared to groups II and III, and was the prevention of depressed birth weight in smokers. In group II, the so-called calorie/protein-balanced complement group, there was a slight, but significant effect on gestational age. In group I (high protein supplementation) was the only significant differences, excess of very early premature births, neonatal deaths, and a significant growth retardation among deliveries up to the thirty-seventh gestational week. The authors postulated that the adverse effects were caused by the intrinsic properties of high protein content. The authors think another possible explanataion is that the great casein content caused a lesser intake of carbohydrates from their staple food, with subsequent changes in the amount of micronutrients absorbed.

In the Danish trial[70] we have taken into account the seasonal variations, which are of great importance, and the pregnant women were matched two and two according to several variables including tobacco consumption.

4. Placental Function and Folic Acid

A positive correlation between birth weight and placental weight has been found in many investigations. The effect of moderate maternal malnutrition on the placenta has been examined.[71] A significant (11 to 15%) reduction in weight was found, but without a change in biochemical compositions of fat, water, protein, DNA, and hydroxyproline.

In the Australian trial[59] of folic acid supplementation, urine estrogens were found to be independent of the folate status. In our prospective study (Denmark),[17] the placental weight was 11.9% (not significant) and the DNA was 11.1% (not significant) higher in the folic acid supplemented group than in the control group. After this we examined, in a double-blind trial, whether folic acid supplement through the whole year could have an influence on the placental function measured by the concentrations of serum placental lactogen hormone (HPL). Seventy-six (76) pregnant women from the antenatal clinic with a possibility of low placental function based on clinical findings and low concentrations of HPL[72] were given a supplement of either folic acid (5 mg per day) or placebo. All were supplemented for at least 6 weeks during pregnancy. Erythrocyte folate was determined weekly during medication. Forty-seven (47) took the medication for 6 weeks or more: 5 of these were excluded, 22 took folic acid, and 20 took a placebo. The average fetal and placental weight, and the maximal increase of HPL were calculated, and the results are seen in Table 2. For all three parameters the values were highest in the folic acid group, namely 5.5% higher for fetal weight, 9.0% higher for placental weight, and 25% higher for Δ HPL. These differences, however, were not statistically significant. The trial was carried out through the whole year, so we found it was not possible to pay attention to the seasonal variations. The supplement of folic acid was given over 6 to 14 weeks, while in the previous study folic acid was given over 20 weeks. These factors are the probable cause of the differences found in the two trials.

VII. VITAMINS AND DEFECTIVE EARLY INTRAUTERINE GROWTH

The fetus is very susceptible in the first few weeks of life to disturbances which lead to congenital malformation or spontaneous abortion. Rubella virus and thalidomide are the most striking examples. It is well known that about one half of the abortions have chromosome

Table 2
DIFFERENCES IN FETAL AND PLACENTAL WEIGHT AND THE MAXIMUM INCREASE IN HPL BETWEEN THE FOLIC ACID-TREATED AND PLACEBO GROUPS

		N	Mean	T-value	Difference (%)	Significance
Fetal weight (g)	+ Folic acid	22	3331	1.71	5.5	$0.1 < p < 0.05$
	− Folic acid	20	3158			
Placental weight (g)	+ Folic acid	21	467	1.33	9.0	N.S.
	− Folic acid	18	429			
ΔHPL mg/ℓ	+ Folic acid	22	2.63	1.16	25	N.S.
	− Folic acid	20	2.10			

Note: N.S. = not significant.

defects, and when that number is compared with the incidence of abortions it will be clear that the majority of chromosome anomalies end in abortion.

Some vitamin deficiencies are known to lead to chromosome anomalies. Lawler et al.[73] found that B_{12} and folic acid deficiency produced anomalies in bone marrow, and that they were reversible within 48 hr. It has been postulated but not proven that other vitamin deficiencies cause such alterations.

A. Vitamin Deficiency and Spontaneous Abortions

About 10% of all pregnancies terminate in spontaneous abortion and about 50% of those have chromosome anomalies, which are postulated to be the cause of the abortions. It is not known how many of the early abortions also have other anomalies. Sandahl[74] found a seasonal variation in the spontaneous abortion rate, which points to the fact that environmental factors are operating.

There is no doubt that severe vitamin deficiency can cause abortion. This was demonstrated in humans by Thiersch,[75] who induced abortions in 10 of 12 cases by giving the folic acid antagonist 4-aminopteroyl glutamic acid. In rats a total lack of folic acid will lead to the resorption of the fetuses.[76]

Martin et al.[77] have measured serum folic acid in 150 women with spontaneous abortions before the eighteenth week of pregnancy and found that 25% had a low level of serum folic acid. They were interpreted as low, however, there was no control group. A similar association was published on several occasions by Hibbard,[63] who found that 15% of spontaneous abortions had low red blood cell folate. In contrast, Chanarin et al.[5] did not find that shortage of folic acid led to spontaneous abortion.

Rolschau, Kristoffersen, and Honoré[78] investigated the red blood cell-folic acid concentration in two groups of abortions, one spontaneous and one induced. The analysis was performed in the spring, when the greatest risk of low folic acid intake is presumed to occur. The two groups were chosen consecutively amongst aborting women without surgical or severe medical disorders. No one had passed the end of the twelfth week. There were 25 in each group, but in the spontaneous abortion group 7 had passed the end of the twelfth week. In the second group all 25 women were included, and they were all attending the hospital for legal abortion (free abortion).

A venous blood sample was drawn from both groups the following morning after fasting. Erythrocyte folic acid was measured with a competitive protein-binding assay. It is seen in Table 3 that erythrocyte folic acid was lower in the group of induced abortions, but the variance was so high that a comparison of the two mean values was not possible. It can be concluded that, as a group, the spontaneous abortions do not have lower folic acid values

Table 3
RED BLOOD CELL FOLIC ACID MEASURED IN
SPONTANEOUS AND INDUCED ABORTIONS

Abortions	No.	Mean RBCFA (ng/mℓ)	S.D.	Gestation (weeks)	Mean age
Spontaneous	18	1710	208	10.6	25.5
Induced	25	1417	108	8.9	27.5

than the induced abortions. It is unclear whether some of them are due to low folic acid values in the erythrocytes. The possibility exists that folic acid deficiency in some women causes an insufficient placentation, as proposed by Martin et al.,[22] or that folic acid deficiency produces chromosome anomalies.[73] It would have been more beneficial if the abortions had been simultaneously examined both for chromosome anomalies and for folic acid status. Another possibility is a bad utilization of folate, possibly due to deficient folic reductase activity in the placenta.

B. Vitamins and Congenital Malformations

The interest in preventing congenital malformations is enormous. Environmental factors such as operating room environment, alcohol, and anticonvulsants are reviewed by Smithells.[79] The vitamins in question are folic acid, B_{12}, and vitamin K. As consequences are serious, even less firm evidence would warrant thorough consideration. The evidence is circumstantial from animal experiments and from trials in humans.

1. Circumstantial Evidence in Congenital Malformations Related to Vitamins

It has become increasingly clear that intrauterine growth retardation is related to congenital malformations. The first could be a result of the latter or they could have a common cause. Wald et al.[80] reported that 20 fetuses with spina bifida were growth retarded. Their biparietal diameter was 0.83-cm smaller than a control group. Another study[81] found a convincing relation between congenital malformations and small-for-date infants. Among small-for-date infants, the incidence of minor malformations was significantly increased. They have a rate of 8.6% for major malformations, 36% for neonatal death, and 14% for stillbirths. In a prospective study of 6376 pregnancies in Sweden, Kullander and Källén[82] found that mothers who had a miscarriage or a stillborn infant used less iron or vitamin preparations in the last half of their pregnancy. In the first half of their pregnancy, the intake was as frequent as that among mothers who gave birth to normal infants. A similar connection to vitamin intake is found by Klemetti.[83]

Indirectly there is support for the association between growth retardation and vitamin deficiency through the work of Baker et al.[40] They compared the concentrations of different vitamins in neonates in two groups consisting of 50 normal-birth-weight neonates and 50 low-birth-weight neonates. The latter group had significantly lower plasma values of B_{12}, folate, and pantothenate. The level of the three vitamins was higher in the infants than in the mothers. It is further seen that these differences were lower for the low-birth-weight infants. It seemed that these low-birth-weight infants received a lesser fraction of their mother's vitamins, possibly because of specific placental insufficiency.

Gross et al.[84] examined 14 African children, whose mothers had been severely folate deficient. Comparing the children with a control group who were iron deficient, developmental disturbances were found in 8 of 14. This is interesting, but the cause has not been proven.

Warfarin acts as a vitamin K antagonist and Stevenson et al.[85] observed the reaction in pregnancy outcome. Administered before the eighth week it often leads to malformations

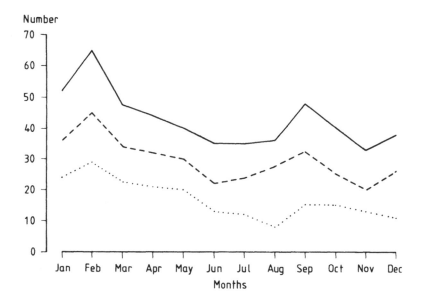

Number

FIGURE 2.   Inheritance of harelip and cleft palate. Ellipses: cleft palate; dotted line = harelip with or without cleft palate; straight line = both curves accumulated. Calculated by and drawn from Fogh-Andersen.[90]

of the face and hands and mental retardation. When warfarin was given in the second midtrimester, it led to opticus atrophy and developmental retardation. It is presumed that this is an effect of vitamin K deficiency and not of warfarin per se. It is generally accepted that epileptic women give birth to a higher incidence of malformed infants. This rate may be due to the mothers' use of anticonvulsants.[86] These seem to cause folic acid deficiency, and it is possible that this is the pathogenesis in the higher malformation rate in epileptic women. Another possibility is a direct teratogenic effect of anticonvulsants.

Blake and Fallinger[87] showed that a phenytoin metabolite found in the urine of newborn infants exposed to phenytoin could be involved in the teratogenic activity of that drug.

Smithells et al.[88] examined vitamin levels in the first trimester of 959 pregnant women. Six CNS-defective infants were born; the mothers had significantly lower levels of serum folate, red cell folate, white blood cell vitamin C, and riboflavin, as compared with the other 953 pregnant women.

Recently Fraser et al.[89] have emphasized a connection between CNS malformations and other malformations. In siblings they found an increased frequency of tracheoesophageal dysraphism, and cleft lip and palate. On the basis of Fogh-Andersen's[90] work, we have constructed a curve to see if there are seasonal variations in the material of cleft lip and palate. There were 480 infants with one or both of these disorders without familiar disposition. The incidences of deliveries for each month during the year were calculated. The time of the formation of the lip and palate were expected to be 34 weeks backward from the time of the delivery and they were referred to that point. The incidences for harelip and cleft palate with or without harelip were added, and it is seen from Figure 2 that a higher proportion of these malformations was grounded in the late winter.

The seasonal variation in malformations are underestimated, and an accumulation of more materials already published could further elucidate the possible environmental role.

2. Animal Experiments

The majority of animal experiments have been done with folic acid on rats. An exception is the observation of Thiersch[75] that a folic acid antagonist produces malformations in humans in a large proportion. Herbert and Tisman[91] have reviewed the evidence supporting an association between B_{12} and folate deficiency and the development of CNS. Marshall-Johnson[92] found decreased activity in mitoses in the neural epithelium when they gave their rats a pteroyl glutamic acid antimetabolite from day 8 to 10 of gestation. In a similar study, hydrocephalus was observed.[76] Woodard and Newberne[93] showed the same result when they gave the rats a B_{12}-free diet.

Skeletal malformations and retarded visceral development in rats are reported by Nelson et al.[94] Anomalies of the great vessels are also reported.[95] Various eye anomalies in rats, depending on the time of the folic acid deficiency in the gestation, are demonstrated by Armstrong and Monie.[96]

The animal experiments on folic acid deficiency demonstrate the profound role of the vitamin, but this is a long way from evaluation of spontaneous malformations in humans.

3. Prospective Studies

Smithells et al.[97] have published materials from which cautious, but important conclusions can be drawn. It concerned a nonrandomized trial among two groups of women who earlier had delivered a neural tube-defect infant. One group was supplied with vitamins more than 28 days before conception; the control group was comprised of women who already were pregnant at their acceptance into the trial, or who refused to participate.

The "supplied" group was given tablets three times a day and the total content per day was vitamin A, 4000 I.U.; vitamin D, 400 I.U.; thiamin, 1.5 mg; riboflavin, 1.5 mg; pyridoxine, 1 mg; nicotinamide, 15 mg; AA, 40 mg; folic acid, 0.36 mg; ferrous sulfate, 75.6 mg; and calcium phosphate 480 mg. One of 188 infants/fetuses in the supplied group had neural tube defect as did 13 of 269 in the control group. These results are impressive, but there are objections to the scientific design, particularly because the women were not appointed to individual groups by the researchers, but selected their preference themselves. After this trial and the trial of Laurence et al.,[98] it is hardly possible to perform a double-blind, randomized trial on each of the vitamins.

Laurence et al.[98] have performed a double-blind, randomized trial with 4 mg folic acid and placebo. One hundred and four (104) women who earlier gave birth to infants with neural tube defect and who planned new pregnancies, were randomized into two groups and given tablets from the day contraception stopped. After 6 to 9 weeks of estimated gestation serum folic acid was measured. Compliance was defined as serum folic acid higher than 10 $\mu g/\ell$. The materials were divided into 3 groups:

	Compliers	Noncompliers	Placebo
Normal fetuses	44	14	47
Fetuses with neural tube defect	0	2	4

Testing compliers against noncompliers plus placebo showed a significantly lower incidence among compliers ($p = 0.04$). The result seems valid, but it is arguable that the limit between compliers and noncompliers is made afterward and fixed arbitrarily. A dietary anamnesis (with regard to folic acid) of the women before conception revealed that the women delivering a neural tube defect fetus had a poor diet.

The incidence of harelip with or without cleft palate has been found to decrease by vitamin supplementation before pregnancy or early in the first trimester. Convincing results have been achieved by Conway[99] and Briggs.[100] Recently, Tolarova[101] has reported that the recurrence rate of cleft lip was significantly reduced in an intervention trial by supplying a combined vitamin preparation with 10 mg of folic acid. Two hundred ninety-seven (297) women participated, but were not randomized. The control group was comprised of women who did not accept the offer of being supplied before conception.

The results of the mentioned investigations are so important that we have to accept them as convincing but not proven.

VIII. PROPOSALS FOR FURTHER INVESTIGATIONS

The three most interesting and promising problems to be solved in the future are

1. Can folic supplement improve the placental function and subsequently minimize the frequency of children born with low birth weight either because of preterm labor or intrauterine growth retardation?
2. Can preconceptional folic acid supplement minimize the incidence of congenital malformations?
3. Is folic acid the only or most essential factor in the above-mentioned problems or are other vitamins implicated in the possible effective mechanisms?

It is obvious from animal experiments and in vitro trials that without folic acid, no cell proliferation can take place. The coincidence of low folate state and children born with neural tube defects or harelip and cleft palate indicate the suspicion of a causal relationship. However, the small incidence of these defects in the population, with the recurrence risk being about 5%, and the fact that there is a hereditary moment, indicate that the pathogenesis must be multifactorial.

To solve the three above-mentioned problems we believe that we may benefit from case-control studies. These studies should be comprised of the measurement of the concentrations of enzymes, amino acids, folic acid compounds, or other vitamins and protein carriers in both maternal and cord blood from cases with congenital malformations and from cases with severe intrauterine growth retardation.

In accordance with this, Baker et al.[45] showed the interesting differences in the ratios between folic acid, B_{12}, and pantothenic acid in maternal blood and cord blood from low birth weight infants as compared with normal weight infants. Futhermore, in vitro investigations on the placenta as concerns the enzymatic function in transformation and transport of folic acid and other vitamins would be worthwhile. The kinetics of folic acid transport and transformation in the placental cells in relation to amino acid states may be interesting, if we bear in mind the formimino transferase deficiency syndrome and the fact that mothers with high concentrations of phenylalanine in blood run a great risk of giving birth to malformed infants.[102]

The future intervention trials with vitamins in a large number of pregnant women must be designed very strictly, since the many variables in the mothers themselves, their surroundings, and their intake, make a conclusion extremely difficult.

REFERENCES

1. **Chanarin, I., Rothman, D., Perry, J., and Stratfull, D.,** Normal dietary folate, iron and protein intake, with particular reference to pregnancy, *Br. Med. J.,* 2, 394, 1968.
2. **Hansen, H. A. and Rybo, G.,** Folsyreprofylax under graviditet, *Nord. Med.,* 76, 867, 1966.
3. **Streiff, R. R. and Little, B.,** Folic acid deficiency in pregnancy, *N. Engl. J. Med.,* 276, 776, 1967.
4. **Avery, B. and Ledger, W. J.,** Folic acid metabolism in well-nourished pregnant women, *Obstet. Gynecol.,* 35, 616, 1970.
5. **Chanarin, I., Rothman, D., Ward, A., and Perry, J.,** Folate status and requirement in pregnancy, *Br. Med. J.,* 2, 390, 1968.

6. **Moscovitch, L. F. and Cooper, B. A.,** Folate content of diets in pregnancy: comparison of diets collected at home and diets prepared from dietary records, *Am. J. Clin. Nutr.,* 26, 707, 1973.

7. **Elsborg, L. and Rosenquist, A.,** Folate intake by teenage girls and by pregnant women, *Int. J. Vitam. Nutr. Res.,* 49, 70, 1979.

8. National Academy of Sciences, Recommended Dietary Allowances, 9th ed., National Academy of Sciences, Washington. D.C., 1980.

9. **Lowenstein, L., Cantlie, G., Ramos, O., and Brunton, L.,** The incidence and prevention of folate deficiency in a pregnant clinic population, *Can. Med. Assoc. J.,* 95, 797, 1966.

10. **Herbert, V. A.,** A palatable diet for producing experimental folate deficiency in man, *Am. J. Clin. Nutr.,* 12, 17, 1963.

11. **Russell, R. M., Ismail-Beigi, F., and Reinhold, J. C.,** Folate content of Iranian breads and the effect of their fiber content on the intestinal absorption of folic acid, *Am. J. Clin. Nutr.,* 29, 799, 1976.

12. **Margo, G., Barker, M., Fernandes-Costa, F., Colman, N., Green, R., and Metz, J.,** Prevention of folate deficiency by food fortification. VII. The use of bread as a vehicle for folate supplementation, *Am. J. Clin. Nutr.,* 28, 761, 1975.

13. **Anderson, B., Belcher, E. H., Chanarin, I., and Mollin, D. L.,** The urinary and faecal excretion of radioactivity after oral doses of 3H-folic acid, *Br. J. Haematol.,* 6, 439, 1960.

14. **Fleming, A. F.,** Seasonal incidence of anemia in pregnancy in Ibadan, *Am. J. Clin. Nutr.,* 23, 224, 1970.

15. **Chanarin, I., MacGibbon, B. M., O'Sullivan, W. J., and Mollin, D. L.,** Folic-acid deficiency in pregnancy. The pathogenesis of megaloblastic anaemia of pregnancy, *Lancet,* 2, 634, 1959.

16. **Yusufji, D., Mathan, V. I., and Baker, S. J.,** Iron, folate and vitamin B_{12} nutrition in pregnancy: a study of 1000 women from southern India, *Bull. WHO,* 48, 15, 1973.

17. **Rolschau, J., Date, J., and Kristoffersen, K.,** Folic acid supplement and uterine growth, *Acta Obstet. Gynecol. Scand.,* 58, 343, 1979.

18. **Thompson, R. B.,** Seasonal incidence of megaloblastic anaemia of pregnancy and the puerperium, *Lancet,* 1, 1171, 1957.

19. **Elsborg, L.,** Pteroyl polyglutamate hydrolase and intestinal absorption of folate polyglutamates, *Dan. Med. Bull.,* 27, 205, 1980.

20. **Waxman, S.,** Folate binding proteins, *Br. J. Haematol.,* 29, 23, 1975.

21. **Waxman, S. and Schreiber, C.,** The purification and characterization of the low molecular weight human folate binding protein using affinity chromatography, *Biochemistry,* 14, 5422, 1975.

22. **Martin, J. D., Davis, R. E., and Stenhouse, J.,** Serum folate and vitamin B_{12} levels in pregnancy with particular reference to uterine bleeding and bacteriuria, *J. Obstet. Gynaecol. Br. Commonw.,* 74, 697, 1967.

23. **Ellegaard, J. and Esmann, V.,** Folate deficiency in malnutrition, malabsorption and during phenytoin treatment diagnosed by determination of serum synthesis in lymphocytes, *Eur. J. Clin. Invest.,* 2, 315, 1972.

24. **Eichner, E. R., Buchanan, B., Smith, J. W., and Hillman, R. S.,** Variations in the hematologic and medical status of alcoholics, *Am. J. Med. Sci.,* 263, 35, 1972.

25. **Alter, H. J., Zvaifler, N. J., and Rath, C. E.,** Interrelationship of rheumatoid arthritis, folic acid and aspirin, *Blood,* 38, 405, 1971.

26. **Reynolds, E. H., Milner, G., Matthews, D. M., and Chanarin, I.,** Anticonvulsants therapy, megaloblastic haemopoiesis and folic acid metabolism, *Q. J. Med.,* 35, 521, 1966.

27. **Gatenby, T. B. B. and Lillie, E. W.,** Clinical analysis of 100 cases of severe megaloblastic anemia of pregnancy, *Br. Med. J.,* 6, 1111, 1960.

28. **Waxman, S. and Schreiber, C.,** The role of folic acid binding proteins (FABP) in the cellular uptake of folates, *Proc. Soc. Exp. Biol. Med.,* 147, 760, 1974.

29. **Erbe, R. W.,** Inborn errors of folate metabolism, *N. Engl. J. Med.,* 293, 753, and 807, 1975.

30. **Arakawa, T.,** Congenital defects in folate utilization, *Am. J. Med.,* 48, 594, 1970.

31. **Rauramo, L., Castren, O., Levanto, A., Markkanen, T., Kivikosko, A., and Ruponen, S.,** Serum folic acid content in pregnancy, *Acta Obstet. Gynecol. Scand. Suppl.,* 7, 101, 1967.

32. **Hansen, H. A. and Rybo, G.,** Folic acid dosage in prophylactic treatment during pregnancy, *Acta Obstet. Gynecol. Scand. Suppl.,* 7, 107, 1967.

33. **Ek, J. and Magnus, E. M.,** Plasma and red blood cell folate during normal pregnancies, *Acta Obstet. Gynecol. Scand.,* 60, 247, 1981.

34. **Colman, N., Larsen, J. V., Barker, M., Barker, E. A., Green, R., and Metz, J.,** Prevention of folate deficiency by food fortification. III. Effect in pregnant subjects of varying amounts of added folic acid, *Am. J. Clin. Nutr.,* 28, 465, 1975.

35. **Markkanen, T., Himanen, P., Pajula, R. L., Ruponen, S., and Castren, O.,** Binding of folic acid to serum proteins. I. The effect of pregnancy, *Acta Haematol.,* 50, 85, 1973.

36. **Areekul, S., Yamarat, P., and Vongyuthithum, M.,** Folic acid and folate binding proteins in pregnancy, *J. Nutr. Sci. Vitaminol.,* 23, 447, 1977.

37. **Fleming, A. F.,** Urinary excretion of folate in pregnancy, *J. Obstet. Gynecol.,* 79, 916, 1972.
38. **Ek, J.,** Plasma and red cell folate values in newborn infants and their mothers in relation to gestational age, *J. Pediatr.,* 97, 288, 1980.
39. **Zachau-Christiansen, B., Hoff-Jørgensen, E., and Østergård Kristensen, H. P.,** The relative haemoglobin, iron, vitamin B_{12} and folic acid values in the blood of mothers and their newborn infants, *Dan. Med. Bull.,* 9, 158, 1962.
40. **Baker, H., Thind, I. S., Frank, O., DeAngelis, B., Caterini, H., and Louria, D. B.,** Vitamin levels in low-birth-weight newborn infants and their mothers, *Am. J. Obstet. Gynecol.,* 129, 521, 1977.
41. **Kamen, B. A. and Caston, J. D.,** Purification of folate binding factor in normal umbilical cord serum, *Proc. Natl. Acad. Sci. U.S.A.,* 72, 4261, 1975.
42. **Kaminetzky, H. A., Baker, H., Frank, O., and Langer, A.,** The effects of intravenously administered water-soluble vitamins during labor in normovitaminemic and hypovitaminemic gravidas on maternal and neonatal blood vitamin levels at delivery, *Am. J. Obstet. Gynecol.,* 120, 697, 1974.
43. **Grundy, M. F. B., Hood, J., and Newman, G. B.,** Birth weight standards in a community of mixed racial origin, *Br. J. Obstet. Gynecol.,* 85, 481, 1978.
44. **Tafari, N., Naeye, R. L., and Gobezie, A.,** Effects of maternal undernutrition and heavy physical work during pregnancy on birth weight, *Br. J. Obstet. Gynecol.,* 87, 222, 1980.
45. **Ulrich, M.,** Fetal growth patterns in a population of Danish newborn infants. II., *Acta Paedr. Scand. Suppl.* 278, 18, 1982.
46. **Billewicz, W. Z. and Thomson, A. M.,** Birth weights in consecutive pregnancies, *J. Obstet. Gynaecol. Br. Commonw.,* 80, 491, 1973.
47. **Winick, M. and Noble, A.,** Quantitative changes in DNA, RNA and protein during prenatal and postnatal growth in the rat, *Dev. Biol.,* 12, 451, 1965.
48. **Oh, W. and Guy, J. A.,** Cellular growth in experimental intrauterine growth retardation in rats, *J. Nutr.,* 101, 1631, 1971.
49. **Krüger, H. and Arias-Stella, J.,** The placenta and the newborn infant at high altitudes, *Am. J. Obstet. Gynecol.,*106, 586, 1970.
50. **Altshuler, G., Russel, P., and Ercomilla, R.,** The placental pathology of small-for-gestational age infants, *Am. J. Obstet. Gynecol.,* 121, 351, 1975.
51. **Rolschau, J.,** Circumvallate placenta and intrauterine growth retardation, *Acta Obstet. Gynecol. Scand. Suppl.,* 72, 11, 1978.
52. **Rosso, P. and Winick, M.,** Intrauterine growth retardation. A new systematic approach based on the clinical and biochemical characteristics of this condition, *J. Perinatal Med.,* 2, 147, 1974.
53. **Rush, D., Stein, Z., and Susser, M., Eds.,** *Diet in Pregnancy: A Randomized Controlled Trial of Nutritional Supplements,* Vol. 16, Alan R. Liss, New York, 1980, chap. 5.
54. **Douglas, J. W. B. and Gear, R.,** Children of low birth weight in the 1946 national cohort, *Arch. Dis. Child.,* 51, 820, 1976.
55. **Fitzhardinge, M. D. and Steven, E. M.,** The small-for-date infant, *Pediatrics,* 49, 671, 1972.
56. **Fancourt, R., Campbell, S., Harvey, D., and Norman, A. P.,** Follow-up study of small-for-dates babies, *Br. Med. J.,* 1, 1435, 1976.
57. **Chase, H. P., Dabière, C. S., Welch, N. N., and O'Brian, D.,** Intrauterine undernutrition and brain development, *Pediatrics,* 47, 491, 1971.
58. **Giles, P. F. H., Harcourt, A. G., and Whiteside, M. G.,** The effect of prescribing folic acid during pregnancy on birth-weight and duration of pregnancy, *Med. J. Austr.,* 2, 17, 1971.
59. **Fleming, A. F., Martin, J. D., Hahnel, R., and Westlake, A. J.,** Effects of iron and folic acid antenatal supplements on maternal haematology and fetal wellbeing, *Med. J. Austr.,* 2, 429, 1974.
60. **Iyengar, L. and Rajalakshmi, K.,** Effects of folic acid supplement on birth weights of infants, *Am. J. Obstet. Gynecol.,* 122, 332, 1975.
61. **Rothman, D.,** Folic acid in pregnancy, *Am. J. Obstet. Gynecol.,* 108, 149, 1970.
62. **Rae, P. G. and Robb, P. M.,** Megaloblastic anemia of pregnancy: a clinical and laboratory study with particular reference to the total and labile serum folate levels, *J. Clin. Pathol.,* 23, 279, 1970.
63. **Hibbard, B. M.,** Folates and the fetus, *S. Afr. Med. J.,* 49, 1223, 1975.
64. **Gandy, G. and Jacobson, W.,** Influence of folic acid on birth-weight and growth of the erythroblastotic infant. I., *Arch. Dis. Child.,* 52, 1, 1977.
65. **Hunt, I. F., Murphy, N. J., Card, B. M., and Jacob, M.,** Dietary folic acid intakes and infant birth weights of pregnant women with low and acceptable red cell folate levels, *Fed. Proc. Fed. Am. Soc. Exp. Biol.,* 38, 711, 1979.
66. **Ek, J.,** Plasma and red cell folate in mothers and infants in normal pregnancies. Relation to birth weight, *Acta Obstet. Gynecol. Scand.,* 61, 17, 1982.
67. **Baumslag, N., Edelstein, T., and Metz, J.,** Reduction of incidence of prematurity by folic acid supplementation in pregnancy, *Br. Med. J.,* 1, 16, 1970.

68. **Whiteside, M. G., Ungar, B., and Cowling, C. D.,** Iron, folic acid and vitamin B_{12} levels in normal pregnancy and their influence on birth weight and the duration of pregnancy, *Med. J. Austr.,* 1, 338, 1968.

69. **Fletcher, J., Gurr, A., Fellingham, F. R., Prankerd, T. A. J., Brant, H. A., and Menzies, D. N.,** The value of folic acid supplements in pregnancy, *J. Obstet. Gynaecol. Br. Commonw.,* 78, 781, 1971.

70. **Rolschau, J.,** A prospective study of the placental weight and content of protein, RNA and DNA, *Acta Obstet. Gynecol. Scand. Suppl.,* 72, 28, 1978.

71. **Lechtig, A., Yarbrough, C., Delgado, H., Martorell, R., Klein, R., and Behar, M.,** Effect of moderate maternal malnutrition on the placenta, *Am. J. Obstet. Gynecol.,* 123, 191, 1975.

72. **Kristoffersen, K., Rolschau, J., Date, J., and Honoré, E.,** The influence of folic acid supplement on intrauterine growth, *9th World Congr. Gynecol. Obstet.,* Abstr. 546, Figo Publishers, Tokyo, 1979.

73. **Lawler, S. D., Roberts, P. D., and Hoffbrand, A. V.,** Chromosome studies in megaloblastic anaemia before and after treatment, *Scand. J. Haematol.,* 8, 309, 1971.

74. **Sandahl, B.,** A study of seasonal and secular trends in incidence of stillbirths and spontaneous abortions in Sweden, *Acta Obstet. Gynecol. Scand.,* 53, 251, 1974.

75. **Thiersch, J. B.,** Therapeutic abortions with a folic acid antagonist 4-aminopteroylglutamic acid (4-amino-PGA) administered by the oral route, *Am. J. Obstet. Gynecol.,* 63, 1298, 1952.

76. **Stempak, J. G.,** Etiology of antenatal hydrocephalus induced by folic acid deficiency in the albino rat, *Anat. Rev.,* 151, 287, 1966.

77. **Martin, R. H., Harper, T. A., and Kelso, W.,** Serum-folic-acid in recurrent abortions, *Lancet,* 1, 670, 1965.

78. **Rolschau, J., Kristoffersen, K., and Honoré, E.,** Erythrocyte folic acid in spontaneous and legal abortions, in press.

79. **Smithells, R. W.,** Environmental teratogens of man, *Br. Med. Bull.,* 32, 27, 1976.

80. **Wald, N., Cuckle, H., and Boreham, J.,** Small biparietal diameter of fetuses with spina bifida: implications for antenatal screening, *Br. J. Obstet. Gynaecol.,* 87, 219, 1980.

81. **Drew, J. H., Parkinson, P., Walstab, J. E., and Beischer, N. A.,** Incidences and types of malformations in newborn infants, *Med. J. Austr.,* 1, 945, 1977.

82. **Kullander, S. and Källén, B.,** A prospective study of drugs and pregnancy, *Acta Obstet. Gynecol. Scand.,* 55, 287, 1976.

83. **Klemetti, A.,** Environmental factors and congenital malformations, a prospective study, *Acta Opthalmol.,* 46, 350, 1968.

84. **Gross, R. L., Newberne, P. M., and Reid, J. V. O.,** Adverse effects on infant development associated with maternal folic acid deficiency, *Nutr. Rep. Int.,* 10, 241, 1974.

85. **Stevenson, R. E., Burton, M., Ferlacto, G. J., and Taylor, H. M.,** Hazards of oral anticoagulants during pregnancy, *JAMA,* 243, 1549, 1980.

86. **Speidel, B. D. and Meadow, S. R.,** Epilepsy, anticonvulsants and congenital malformations, *Drugs,* 8, 354, 1974.

87. **Blake, D. A. and Fallinger, C.,** Embryopathic interaction of phenytoin and trichloropropene oxide in mice, *Teratology,* 13, 17A, 1976.

88. **Smithells, R. W., Sheppard, S., and Schorah, C. J.,** Vitamin deficiencies and neural tube defects, *Arch. Dis. Child.,* 51, 944, 1976.

89. **Fraser, F. C., Czeizel, A., and Hanson, C.,** Increased frequency of neural tube defects in sibs of children with other malformations, *Lancet,* 2, 144, 1982.

90. **Fogh-Andersen, P.,** *Inheritance of Harelip and Cleft Palate,* Arnold Busch, Copenhagen, 1942.

91. **Herbert, V. and Tisman, G.,** Effects of deficiencies of folic acid and vitamin B_{12} on central nervous system function and development, in *Biology of Brain Dysfunction,* Vol. 1, Gaull, G. E., Ed., Plenum Press, New York, 1973, chap. 10.

92. **Marshall Johnson, E.,** Effects of maternal folic acid deficiency on cytologic phenomena in rat embryo, *Anat. Rev.,* 149, 49, 1964.

93. **Woodard, J. C. and Newberne, P. M.,** The pathogenesis of hydrocephalus in newborn rats deficient in vitamin B_{12}, *J. Embryol. Exp. Morphol.,* 17, 177, 1967.

94. **Nelson, M. M., Asling, C. W., and Evans, H. M.,** Production of multiple congenital abnormalities in young by maternal pteroylglutamic acid deficiency during gestation, *J. Nutr.,* 48, 61, 1952.

95. **Monie, I. W. and Nelson, M. M.,** Abnormalities of pulmonary and other vessels in rat fetuses from maternal pteroylglutamic acid deficiency, *Anat. Rev.,* 147, 397, 1963.

96. **Armstrong, R. C. and Monie, I. W.,** Congenital eye defects following maternal folic-acid deficiency during pregnancy, *J. Embryol. Exp. Morphol.,* 16, 531, 1966.

97. **Smithells, R. W., Sheppard, S., Schorah, C. J., Seller, M. J., Nevin, N. C., Harris, R., Read, A. P., and Fielding, D. W.,** Possible prevention of neural-tube defects by periconceptional vitamin supplementation, *Lancet,* 1, 339, 1980.

98. **Laurence, K. M., James, N., Miller, M. H., Tennant, G. B., and Campbell, H.,** Double-blind randomized controlled trial of folate treatment before conception to prevent recurrence of neural-tube defects, *Br. Med. J.,* 283, 269, 1981.

99. **Conway, H.,** Effect of supplemental vitamin therapy on the limitation of incidence of cleft lip and cleft palate in humans, *Plast. Reconstr. Surg.,* 22, 450, 1958.

100. **Briggs, R. M.,** Vitamin supplementation as a possible factor in the incidence of cleft lip/palate deformities in humans, *Clin. Plast. Surg.,* 3, 647, 1976.

101. **Tolarova, M.,** Periconceptional supplementation with vitamins and folic acid to prevent recurrence of cleft lip, *Lancet,* 2, 217, 1982.

102. **Nielsen, K. B. and Wamberg, E.,** Maternal phenylketonuria. Følling's disease during pregnancy as the cause of congenital microencephaly and mental retardation, *Ugeskr. Laeg.,* 141, 3218, 1979.

Chapter 5

MODIFICATION OF NEOPLASTIC PROCESSES BY VITAMIN A AND RETINOIDS

Jill M. Blunck

TABLE OF CONTENTS

I. INTRODUCTION

The discovery,[1] tissue effects,[2] and structure activity relationships[3] of retinoids (vitamin A is a generic name for all compounds other than carotenoids that qualitatively exhibit the biological activity of retinol[4] — retinoids are analogues of vitamin A) have been amply dealt with in other sources, as have the effects of states of deficiency and excess. Deficiency of vitamin A is an urgent problem worldwide, as described in a recent WHO Technical Report.[5]

According to the "prophets of doom," the exponential increase in the production of new chemicals since the end of World War II includes many unsuspected chemical carcinogens and we can expect a significant increase in cancer rates in the forseeable future.[6] The undoubted success of the medical profession in curing or preventing other significant contributors to human mortality will also increase the population at risk.[7] Depending on how one views the statistics, the proposed "cancer epidemic" may have already begun.[6,8]

Another public health problem that was recently discussed in a WHO Technical Report[5] is the widespread incidence of blindness, particularly in "third world" countries, due to a dietary inadequacy of vitamin A.

While the problem of carcinogens is more acute in the developed nations, in contrast to the third world, the recent trend of exporting hazardous processes to third world countries with their less rigorous public health and occupational hygiene legislation and large, cheap labor forces,[9] will expose populations with marginal vitamin A intakes to many potential carcinogens. The increasing use of fertilizers and pesticides to increase food production in the third world will also increase the potential for carcinogen exposure in these countries.

What are the implications of these disparate pieces of information? Over the past 50 years, an increasing body of evidence derived from studies of man and experimental animals suggests that vitamin A status is an important determinant of one's susceptibility to cancer and the progression of established cancers.[10] The objective of this review is not to cover all of the burgeoning literature on this topic, which has been capably reviewed elsewhere,[3,11-17] but to amplify salient features and cover some of the more recent literature.

II. METABOLISM AND TISSUE DISTRIBUTION OF VITAMIN A

Animal species lack the biochemical pathways necessary to synthesize the carotenoid structure,[18] a precursor of vitamin A. Vitamin A is a generic term for compounds with the biological activity of retinol and was initially called fat soluble A,[4] as it is absorbed in the small intestine with lipids. Carotenoids are likewise absorbed and are split in the intestinal mucosa[4] into two molecules of retinal which is then reduced to retinol.[4,19] Retinol reductase is a zinc-dependent enzyme and there is evidence of reduced conversion of β-carotene to retinol in zinc deficiency.[20] Retinol is then re-esterified with unsaturated fatty acids and transported in chylomicrons to the liver. At the liver, the ester linkage is hydrolyzed and the molecule enters the liver cells as retinol, where it is re-esterified[21] and stored associated with lipid droplets.[4] The esterification of retinol is analogous to that of other cellular lipids, such as cholesterol, and is catalyzed by a fatty acyl coenzyme A retinol acyl transferase.[21] Much retinol (19% of liver vitamin A) is found in a high molecular weight lipoprotein complex in rat liver cytosol; this complex also contains the cytosol — retinol-binding protein, retinyl palmitate hydrolase as well as triolein, cholesterol oleate, and dipalmitoyl phosphatidylcholine hydrolase activities.[22] The complex was isolated by gel filtration and ultracentrifugation, but it may represent an artifact generated during tissue homogenization.[22] The principal liver cell involved in storage of retinol is the Ito[5] or stellate cell which can be found in the space of Disse adjacent to the sinusoids. If retinol is required by the peripheral tissues, it is transported by a retinol-binding protein, which is synthesized in the liver and migrates with the serum globulins; this protein associates with prealbumin, presumably to form a larger molecule that is not filtered by the glomerulus. The retinol-binding protein concentration in plasma and plasma retinol is increased in chronic renal disease[23] and estimation of retinol-binding protein in urine has recently been proposed as a means of detecting disturbed renal tubular function.[24] Retinol-binding protein interacts with specific cell surface receptors at the retinol requiring peripheral tissues.[4] Free retinol exerts a detergent-like action on membranes and is responsible for vitamin A toxicity effects.

The synthesis and secretion of retinol-binding protein is regulated by vitamin A status, which is decreased in deficiency states,[4] and by factors that alter liver protein synthesis. There is evidence that retinol is metabolized to retinoic acid in several species; retinoic acid has all the biological properties of retinol except that it will not support vision and reproductive functions.[25] However, recent evidence suggests that retinoic acid stimulates testosterone production by Leydig cells in the testis. There is some in vivo isomerization of *all-trans* retinoic acid to 13-*cis*-retinoic acid, which is further metabolized to more polar derivatives such as 5,6-epoxy retinoic acid[4] and 4-hydroxy-and 4-oxo derivatives in various rat tissues,[26] including liver.[27] There is a retinoic acid 5,6-epoxidase in rat kidney microsomes which is dependent on ATP, NADPH, and O_2. It is stimulated by Fe^{2+} and has properties like microsomal lipid peroxidases.[28] If epoxidation is blocked, the function of retinoic acid is not inhibited, suggesting that epoxidation is not a necessary metabolic step for biological activity.[28] The reaction is probably a nonspecific lipid peroxidation reaction. The 4-oxo and 5,6-epoxy metabolites appear to have less biological activity than retinoic acid.[29] However, there are multiple metabolites which are at least partially tissue specific[30] and retinoids induce retinoic acid metabolism in vitamin A-deficient hamsters.[31] Oxidation of retinoic acid is believed to be mediated by a cytochrome P_{450}-type enzyme system.[32] However, retinoyl-β-glucuronide, a major metabolite of retinoic acid[19] in the small intestine[33] that is also produced by incubating 5,6-epoxy retinoic acid with uridine diphospho 1α-D-glucuronic acid in the presence of liver microsomes,[34] has increased biological activity and may be either an active metabolite or a tolerated modification. The suggestion has been made that glucuronidation ensures conservation of vitamin A in the body by enterohepatic circulation;[35] retinol, retinoic acid and 13-*cis*-retinoic acid are likewise excreted in the bile as β-glucuronides.[34] Ethanol administration increased the metabolism of *all-trans*-11,12-^{3}H retinoic acid to 4-hydroxylated derivatives but had no effect on β-glucuronidation in the presence of uridine diphosphoglucuronic acid (UDPGA).[27] As the 4-hydroxylation reaction may represent a catabolic pathway, this could account for the decrease in liver vitamin A in rats consuming ethanol.[27] In human skin extracts, the major retinoids are *all-trans*-retinol and 3,4-dehydroretinol, the latter compound being found in dyskeratotic skin where its further metabolism may be defective.[36]

Liver stores of vitamin A are probably the best index of vitamin A status;[37,38] plasma levels of vitamin A show little correlation with dietary intake and reflect the amount of plasma retinol-binding protein. Plasma levels may decline when the liver stores are exhausted. If a dose of radioactively labelled retinol is given to a deficient rat, there is an increase in the excretion of labelled metabolites in urine in comparison with an animal of normal vitamin A status.[37] This is because incoming retinol in such circumstances is preferentially utilized by the peripheral tissues rather than stored. The minimum liver concentration of retinol necessary to favor storage rather than immediate turnover and breakdown is 60 IU/g liver.[38] When animals or humans are subjected to a variety of stressful stimuli, there is a decrease in serum vitamin A, although there is no change in the amount of the vitamin in the liver or testis. There is evidence that the low serum vitamin A is due to increased renal loss in these circumstances.[39] During stress there is an accumulation of vitamin A in the enlarged adrenal glands. The suggestion that vitamin A plays an essential role in steroid hormone dynamics is borne out by the fact that vitamin A accumulates preferentially in the adrenal glands when the vitamin is first administered to animals depleted of vitamin A.[39]

Interestingly, liver vitamin A stores are very low in alcoholics even with mild liver damage, even though plasma vitamin A levels are often normal;[40] this has some possible bearing on the increased risk of hepatic neoplasms in persons with alcoholic cirrhosis.

Both uptake and liver storage of vitamin A are equally effective when the vitamin is administered in an aqueous suspension or an oily solution,[41] thus administration of vitamins A and E in solutions given for total parenteral nutrition (TPN) maintains blood levels and leads to storage.[41] Illness, infections, and prematurity affect both the storage and tissue requirements of vitamins A and E.[41]

III. RETINOL-BINDING PROTEIN

Retinol-binding protein (RBP) migrates with the serum globulins and is synthesized in the liver; serum levels are often decreased in liver disease.[42] More research is needed into factors regulating the synthesis and secretion of RBP.[43] RBP has been variously described as a protein, first isolated in 1968,[4] of molecular weight 67,000 sedimenting at 4.6S[44] or as a protein of molecular weight 20,000[45] which circulates in the bloodstream as a 1:1 complex with serum prealbumin;[42] this prevents loss by glomerular filtration. In fact, the kidney is the principal site of RBP degradation, and serum RBP levels are reported to be elevated in renal disease.[23,42] RBP can cross the placental barrier. Various factors significantly modify RBP synthesis and secretion. For example, if the diet is protein deficient, there is decreased apoRBP production and consequently a decreased plasma retinol.[46] Zinc deficiency is associated with decreased plasma retinol and the liver content of vitamin A is increased in zinc-deficient rats.[47] Zinc is essential for protein synthesis and may interfere with the synthesis of RBP. Zinc deficiency can cause abnormal dark adaptation and night blindness in individuals with adequate vitamin A status. The effects of zinc supplementation on plasma levels of vitamin A and RBP in children being treated for protein-energy malnutrition was investigated by Mathias.[48] If intakes of vitamin A, protein, and energy were similar, there was a sustained increase in plasma vitamin A and RBP if a zinc supplement was given; such increases were not sustained in the absence of zinc supplementation.[48] These results confirm the necessity of zinc for RBP synthesis.

Steroids,[49] including dexamethasone and oral contraceptive agents[50] increase the synthesis of RBP. Female rats administered oral 0.5 mg norgestrel and 0.05 mg ethynylestradiol at 50 times the human dose showed an increase in plasma vitamin A and a decrease in liver vitamin A.[50] There was no marked change in absorption, storage, or metabolism of the vitamin, therefore the conclusion was made that there was no justification for an upward revision of vitamin A requirements in users of oral contraceptives. This increase in plasma vitamin A has been considered to pose a risk of malformed offspring to oral contraceptive users who unwittingly continue to take these drugs in early pregnancy. The effect of chronic immobilization stress on the tissue distribution of vitamin A was investigated in rats fed adequate levels of vitamin A.[39] There was a decrease in serum vitamin A levels and an increase in vitamin A levels in the adrenals; the decreased plasma vitamin A was considered to be due to increased renal loss.[39] In contrast, both steroids and adrenalin increase serum vitamin A.[51] Probably most important, vitamin A stimulates apoRBP synthesis and secretion,[45] and this is the basis of the relative dose response test of liver vitamin A stores.[46] A dose of retinol is given and if liver stores are adequate, there is an increase in plasma vitamin A as holoRBP is released from the liver. A corollary is that mobilization of vitamin A from the liver is decreased by retinol deficiency, and both RBP and prealbumin increase in the liver in this condition. Vitamin A is stored predominantly as retinyl esters in a high molecular weight lipoprotein complex in hepatocytes and there is evidence that retinyl ester hydrolase is part of the complex.[22] The hydrolase activity is enhanced by serum albumin, which binds released retinol and acts as a "sink".[22] The hydrolyzed retinol is transferred to apoRBP in the liver, where the linkage of holoRBP (four molecules) and prealbumin also takes place. Secretion of the complex involves the Golgi apparatus, and probably also the microtubules, for colchichine blocks secretion.[52]

Rats given excess vitamin A show decreases in free and assembled tubulin.[52] This finding could partly explain the hepatotoxicity of excess vitamin A. Prealbumin also transports T_4 and it is interesting that there is an inverse correlation between serum T_4 and vitamin A levels.[54]

There are specific cell surface receptors for RBP in tissues which are targets for vitamin A.[42] If the vitamin A circulates free it is not specifically delivered to tissues and produces toxic effects,[45,55,56] possibly by disrupting cell membranes and those of organelles such as lysosomes. The free vitamin A is possibly in association with serum LDL, as there is a correlation between plasma vitamin A and cholesterol levels.[57] In fact, vitamin A increases serum triglyceride levels in the rat, which is probably due to increased secretion.[58] LDL and holoRBP possibly share determinants for secretion.

Antibodies to cellular retinol-binding protein (cRBP) do not cross react immunologically with serum RBP or cellular retinoic acid-binding protein (cRABP).[59] There is no known condition where apoRBP is absent or has abnormal immunological properties.[42]

The nature of the binding site on RBP for retinol has been investigated.[60] Binding of retinol was not affected by acetylation of lysine residues in RBP, nor by modification of one of eight tyrosine or two of four tryptophan residues. However, alkylation of -S-S- groups disrupts binding. Retinoid binding is not very sensitive to configurational changes in the polyene side chain, but the cyclohexene ring is probably necessary. Recently some radioactive retinoid bromoacetates have been synthesized as prospective affinity labels.[60] Binding of such a label followed by proteolytic cleavage and peptide mapping of labelled RBP would identify the amino acids and peptides of the retinoid-binding site on RBP. Thus far, β-11-^{3}H ionylidenethylbromoacetate (IEBA) seems to be the best candidate for an affinity label for RBP; radioactive IEBA is not displaced from RBP by excess retinol, binding of retinol to RBP is decreased to an extent equal to the amount of IEBA that is not displaced from RBP by retinol, and the IEBA-RBP complex is such that IEBA cannot be removed from it by total lipid extraction. The IEBA-RBP complex forms slowly and is covalent, possibly due to reaction of the methylene carbon of the bromoacetyl group of IEBA with the amino acid in RBP that normally interacts with the hydroxyl group of retinol.[60] IEBA may even be used as an affinity label for cRBP, but this has yet to be investigated.

IV. VITAMIN A TOXICITY

Animals cannot synthesize the carotenoid skeleton, therefore a dietary source of vitamin A or carotene is essential.[18] However, vitamin A and retinoids are toxic when administered in large doses,[45,55,56] which may limit their use as agents for the treatment of tumors as they must be used for this purpose in doses approaching toxic doses.[61] However, they may not need to be administered in such large doses for chemoprevention and treatment of skin lesions. By contrast, β-carotene, the retinol precursor molecule in foods of plant origin, is nontoxic and may therefore be more useful for chemoprevention of tumors.[43]

Initial reports on toxicity involved Arctic and Antarctic explorers who ate large quantities of polar bear and seal liver. Although there are no reports of fatal toxicity in man, a specimen of *Homo erectus* showed skeletal evidence of hypervitaminosis A, possibly because of a diet containing carnivore liver.[62] Carnivore liver contains more vitamin A than that of herbivores; man is similar to herbivores.

Retinol has a chain of alternating single and double bonds characterized by high electron mobility. Vitamin A or a metabolite may be present in the lipids of biological membranes and may function in mitochrondrial electron transport.[18]

There is a general consensus that in order for vitamin A toxicity to occur, retinol must be present in a free form.[42,49,55] Thus serum RBP and cRBP, by binding retinol and preventing nonspecific uptake of retinoids by membranes, prevent symptoms of toxicity. This finding is well documented for serum RBP, but such a role has also been suggested for cRBP.[59]

Retinol decreases the microviscosity (increases the fluidity) of cellular membranes[63] and increases the surface area of and hemolyses erythrocytes,[18] although increases in microviscosity occur at higher concentrations. The effects on microviscosity are prevented by d-α-tocopherolacetate, which inhibits vitamin A toxicity.[63] The order of effectiveness of retinoids on microviscosity parallels their order of toxicity in vivo. Retinol may be responsible for labilization of lysosomes and it induces the release of lysosomal acid proteases that are responsible for degrading cartilage matrix.[64] The effects on lysosomes are antagonized by glucocorticoids.[18] Retinol has a synergistic effect with aflatoxin B_1 in inducing hemopericardium in the chick which is prevented by an increase in dietary α-tocopherol.[65] Other effects of hypervitaminosis A are skeletal changes,[56,62,66] fatty liver and liver damage, and effects on CNS.[3] When administered to monkeys (*Macacus fascicularis*) intramuscularly as a single dose. With vitamin D_2 and E (which may decrease the toxicity of vitamin A), 200 mg retinol per kilogram produced early signs of toxicity (yawning, drowsiness, nausea and vomiting, head shaking, neck hyperextension, motor hyperactivity, and incoordination) within 3 to 35 min.[67] After 1 to 2 hr, appetite decreased and the skin became itchy. After 1 to 6 days, weakness, weight loss, and diminished reflexes were apparent. The animals then became comatose and died in respiratory failure, which was sometimes preceded by convulsions.[67] Both the time of onset of symptoms and signs and the survival time after dosing were inversely proportional to the dose. The LD_{50} was calculated as 168 mg retinol per kilogram, which is high.[67] Apparently rats are even less sensitive to hypervitaminosis A, although humans may be more sensitive. In man, hypervitaminosis A induces skin peeling, vomiting, diarrhea, headache, and convulsions.[62] Toxic levels of retinoids may induce lipid deacylation and prostaglandin production,[13] for some toxic effects can be prevented in vivo by administration of inhibitors of prostaglandin synthesis, such as aspirin and other nonsteroid anti-inflammatory drugs.[11]

V. CYTOSOL—RETINOL-BINDING PROTEIN AND CYTOSOL—RETINOIC ACID-BINDING PROTEIN

The means by which retinoids modify cellular functions are unclear; they may act as immunomodulators or exert their effects as coenzymes in glycosylation reactions, modifying cell surface glycoproteins, but there is other evidence that suggests that they may modify cellular metabolism in a manner analogous to steroid hormones.[1,59] In fact, cellular-binding proteins for both retinol, cRBP, and retinoic acid (cRAPB) exist.[44] cRABP is smaller than the proteins that bind steroids (2S as opposed to 4—8S);[69,70] there are 0.5 to 1.5 × 10⁶ molecules of cRBP per liver cell.[70,71] A protein with similar characteristics has recently been isolated from human liver.[72] Retinol does not compete with retinoic acid for binding to cRABP; retinoic acid does not compete with retinol for binding to cRBP.[69] Mouse and hamster cRBP have similar antigenic determinants to those of cRBP in the rat, similar to that in human liver, but retinoid-binding proteins in other species have significantly less or no similarity in antigenicity.[73]

The reaction of the binding proteins with retinoids in vitro is temperature-dependent, saturable, and proportional to the concentration of free retinoid present in the cytosol.[74,75] The function of these proteins may be as intracellular transport proteins, as a vehicle for sequestering retinoids and preventing toxicity, or as a specific protein for transferring retinoids to the nucleus in a role analagous to that of steroid hormone receptors.[59] The nature of the terminal polar group of the retinoid is an important determinant of retinoid binding, thus retinol will not bind to cRABP and retinoic acid will not bind to cRBP. This suggests that both retinol and retinoic acid are "active" forms of vitamin A.[74] The retinoid binding site on the proteins contains sulfydryl groups, for binding is inhibited in vitro by mercurials and the effect is reversed by dithiothreitol.[76] Recently a variety of synthetic polyprenoids have been tested for binding to cRABP from rat testis.[70] A suitable chain length and double bond arrangement were necessary for binding; the ring structure did not affect binding.[70] Dephosphorylation-phosphorylation reactions of cRABP influence the binding capacity for retinoic acid, as dephosphorylation of cRABP from Lewis lung tumor cytosol by alkaline phosphatase treatment enhanced the binding of retinoic acid.[77] An immunoassay has recently been developed for cRBP by Adachi et al.,[59] which can detect 10 to 100 µg of binding protein (either as apo- or as holo-RBP) per assay tube. There was no cross reaction with serum RBP or cRABP. cRBP is present in highest concentrations in kidney, liver, small intestine, lung, spleen, eye, and testis and there is very little in serum, brain, muscle, fat, and heart and surprisingly, in skin and bladder.[59] Detergent and proteolytic enzyme inhibitor addition is necessary in order to assay total cRBP, suggesting that cRBP is associated with intracellular lipid structures.[59] In fact, 95% is found in a lipid-protein aggregate with lipid hydrolases in rat liver cytosol.[22] It is present in cytosol and in the particulate fraction of the cell (microsomes, nuclei, and mitochondria) in testis.[68] This assay is superior to previous assays as it is independent of the degree of saturation of cRBP and reveals 10 to 20 times the amount of cRBP as previous assays.[68] Both holo-cRBP and holo-cRABP react with nuclei in vitro in a specific, saturable, and temperature-dependent manner and transfer retinoids to the nuclei.[71,75] It is not known whether cRBP delivers retinol to the nucleus or translocates with it into the nuclei although studies with labelled cRBP show that it does not bind to nuclei or chromatin unless retinol is attached.[68] Total saturation of cRABP is however not necessary for nuclear uptake of the complex. The nuclear complex sediments at 2S and retinoic acid is loosely bound, dissociating on dialysis. Retinoids interact with binding sites in chromatin which are pronase- and DNAase-, but not RNAase-sensitive,[78] although in HeLa cells, much retinoid is associated with three nuclear membrane proteins. This probably represents free retinol, which binds to the nuclear periphery (probably the membrane).[68] The number of nuclear retinoid binding sites is greater than the number of steroid binding sites by an order of magnitude.[79] This possibly reflects the smaller size of the retinoid carrier proteins. Estimates of the number of retinol molecules bound per nucleus vary from 1.34×10^5 to 8×10^5 molecules.[68] Retinoids must modify gene expression, for they modify the state of cell differentiation. The number of nuclear binding sites is affected by vitamin A status, being greater in vitamin A deficiency states.[52,68,71,74] Retinol must be complexed to protein for binding to nuclei to occur; mitochondria can also take up retinol from cRBP.[68] The amino acid sequences of cRBP and cRABP have some areas of homology but the amino acid composition and N-terminal sequences of the first 16 amino acids of cRBP differs from RBP.[72] Nevertheless, antibodies to cRBP, cRABP, and serum RBP do not cross-react.[59] There is also some amino acid sequence homology with a myelin protein,[80] which is not surprising in view of the affinity of these proteins for membrane-bound structures.

The quantity of these proteins in the cell is influenced by vitamin A status, being greater in vitamin A deficiency,[52,68,71,74] and also varies during development.[81] cRABP seems to be preferentially present in embryonic and undifferentiated tissues. In fact, cRABP has been reported to be present in increased concentrations in certain tumors as compared with the tissue of origin.[82] There is a special type of binding protein, F type, which is an oncofetal protein in hepatocellular carcinoma which is absent in adult liver.[83] The protein is also present in human fetal liver and rat hyperplastic liver.[83] In rat mammary tissue, the level of cRABP is increased by administering estradiol-17β, but not by progesterone.[84] There are several reports that the amount of these proteins present in tissues determines the biological response of the particular tissue to retinoids.[49,69,74] Assays for these proteins may be a means of detecting those tumors likely to be sensitive to retinoid therapy. However, a recent study of various tumor cell lines which differed in sensitivity to retinoids indicated that the content of retinoid-binding proteins did not significantly vary from one cell type to another and thus were not responsible for the differences in cellular response to retinoids.[85] In contrast, Mehta et al.[86] reported a correlation between the inhibition of *N*-methyl-*N*-nitroso urea (MNU)-induced mammary carcinogenesis and levels of cRABP. cRABP was increased following ovariectomy when retinoid suppression of carcinogenesis was more effective.[86] However, later investigations revealed that there was no correlation between cRABP and steroid receptors in human breast cancer tissue. cRABP amounts were increased in the better differentiated tumors, which may be more sensitive to retinoid therapy.[86] Bollag and Matter[87] concluded that binding proteins are necessary, but not sufficient prerequisites for retinoid action. Other mechanisms of retinoid action on cell growth certainly must exist, for retinoic acid inhibits growth of human skin fibroblasts, where there may be some cRABP, but no cRBP.

In cultured human epidermal cells, retinol certainly affects gene expression, for in the presence of retinol, synthesis of a 67 Kd keratin is suppressed, while that of 40 and 52 Kd keratins is stimulated. These proteins are all translated from different messenger RNAs.[68]

VI. EFFECTS OF VITAMIN A ON CELL DIVISION

The effects of vitamin A and retinoids on cell division vary. Generally, physiological doses of vitamin A stimulate cell division, while pharmacological/toxicological doses inhibit division. Cell division can also be stimulated in epithelia during vitamin A deficiency.[15] The effect observed varies depending on the dose, the duration of treatment,[88] and type of retinoid used, and the cell or tissue type to which it is applied. Thus the growth of many tumor cell lines is inhibited by retinoids,[89,91] while that of normal cell lines and tissues is stimulated,[75,92-95] sometimes in excess of effects induced by normal stimulating factors such as erythropoietin[96] and colony-stimulating factor.[97] There is a requirement for a free carboxyl group at C15 for optimal inhibition of tumor cell growth by retinoids. Possibly this difference in cell response reflects differences in the content of cRBP between tumor and normal tissues, as the cellular content of these proteins has been suggested to be correlated with retinoid effects,[74] although this is not necessarily so in a variety of neoplastic cell lines.[85]

However, the growth of some normal cells/tissues is also inhibited by retinoids, notably human keratinocytes[99] and rat kidney cells stimulated by growth factors (*vide infra*), although mitotic activity in mouse skin is by contrast, enhanced.[100] There is also decreased layering and adhesion of epidermal cells both in culture and in the skin[101] in the presence of retinoids, which is probably a consequence of the decrease in desmosomes and disruption of microtubules induced by these compounds. By contrast, there is a report of increased DNA synthesis and mitosis in epithelia during vitamin A deficiency.[15]

As the mechanisms regulating growth in normal tissues are imperfectly understood, the mechanism by which retinoids might exert these effects is unclear. However, according to Otto et al.,[102] replication is controlled by the interaction of cells with a variety of growth factors, hormones, nutrients, ions, and other defined compounds, as well as by the spatial relationship among cells. The growth factors are polypeptide hormones which interact with specific cell surface receptors and are internalized, signalling by some unknown means an increase in DNA synthesis and mitosis.[102] Many of these factors stimulate growth optimally if administered in the correct temporal sequence or with specific combinations of other factors.[103] Cells "transformed" (to the malignant phenotype) by Moloney Sarcoma Virus (MSV) produce a sarcoma growth factor which interacts with membrane receptors that are coupled with protein kinases that catalyze the phosphorylation of tyrosine residues in target proteins, including those of the receptor itself.[104] Epidermal growth factor (EGF, urogastrone) is produced by Brunner's glands in the duodenum and by salivary glands,[105] and is likewise coupled with a protein kinase. Tumor-promoting agents interfere with the binding of EGF to its receptor,[106,107] and there are in fact, specific cell surface receptors for the terpene tumor promoters on susceptible cell types.[108] An explanation for this puzzling finding is that terpenes are physiological regulators in primitive life forms such as yeasts;[109] presumably the ability to respond to these compounds has not been lost during evolution to mammalian organisms.

Retinoids in general antagonize the cellular effects of tumor promoters and even some of those induced by growth factors, including sarcoma growth factor.[13,110,111] EGF can in fact act as a tumor promoter.[112] In spite of decreasing the growth of many tumor lines,[98] retinoids generally increase the number of EGF receptors on these cells as well as in normal fibroblasts and epidermal lines.[113,114] Possibly this represents an adaptive cellular response to the decreased cell division induced by retinoids and may explain the enhancement by retinoids of tumor promoter-induced mitogenesis.[115]

Retinoids also interfere with the mediation of sarcoma growth factor-induced cellular effects, such as an increase in plasminogen activator production and transformation of fibroblasts in culture.[13,110,111] An interesting synthetic retinoid-terpene compound has been synthesized, phorbol-12-retinoate-13-acetate (PRA) which combines chemical features of both the retinoids and the phorbol ester tumor promoters.[116] This compound stimulates DNA synthesis and mitosis in the skin of NMRI mice, but is not a tumor promoter itself, although it greatly enhances the tumor yield if given together with one to four applications of the classical mouse skin tumor promoter phorbol myristate acetate (PMA: tetradecanoyl phorbol-13-acetate; TPA). On the basis of this finding, tumor promotion in mouse skin was divided into two stages and PRA was described as a pure second-stage promoting agent.[116]

Some clarification of the apparently opposing effects of retinoids on cell division comes from a recent paper by Schroder et al.,[88] who showed that micromolar retinoic acid potentiates the mitogenic response of murine 3T3 cells to TPA, but if the same cells are exposed to the retinoid for a longer period, there is inhibition of replication after a minimum of 24 hr of treatment, due to decreased passage through the S phase as a result of decreased DNA replication.

VII. EFFECTS OF VITAMIN A ON MORPHOGENESIS

Hypervitaminosis A has long been known to be a teratogen[117] and recent findings indicate that vitamin A exerts profound effects on cell differentiation[118] and limb regeneration in amphibians.[119]

If applied to chick embryos, retinoic acid induces differentiation from scales to feathers on the feet[118] and treatment of F9 embryonic teratocarcinoma cells, which are similar to embryonic ectoderm, with retinoic acid induces the formation of parietal endoderm as judged by absence of α-fetoprotein, production of plasminogen activator, laminin and Type IV collagen, and decreased alkaline phosphatase and lactic dehydrogenase activities.[120,121] The final phenotype of the differentiated cell was not dependent upon the duration of treatment, although the total number of differentiated cells was. Presumably responsive cells responded in an "on/off" manner possibly because of their content of cRBP. The change can be induced by sequential treatment with retinoid and then dibutyryl cAMP.[121] Other work has shown that retinoic acid treatment of cultures induced both plasma membrane-associated and cytosol cAMP-dependent protein kinase activities. There was a preferential increase of regulatory subunit II in the membrane fraction, while regulatory subunit I was increased in the cytosol fraction. It was concluded that the increased cAMP-dependent protein kinase activity and particularly that of regulatory subunit II may be early events in retinoic acid-mediated cellular differentiation. The changes in phosphorylation patterns resulting from altered kinase activity may be responsible for the altered cell morphology following retinoic acid treatment of embryonal carcinoma cells.[122] The induction of protein kinase by retinoic acid thus explains the requirement for sequential treatment with retinoic acid and dibutyryl cAMP to induce parietal endoderm.[121] Once induced, the new cell type is not dependent upon the continued presence of either compound.[121]

If retinoic acid was administered to 129/SV mice with F9 embryonal cell implants, tumor growth was retarded and the survival time of the mice increased, implying that the cells had differentiated.[120]

As well as influencing the differentiation of individual cells, retinoids have profound effects on pattern formation in regenerating tissues. Axolotls are resistant to toxic levels of retinoids, which slow the growth of their limb buds and induce duplication of regenerating limbs.[119] Likewise, if amputations are performed on the limbs of *Bufo* tadpoles and the amputation stumps are treated with filter paper impregnated with retinoic acid or other retinoids, complete limbs can be regenerated following amputation through the carpals.[119] Vitamin A seems to have an effect on limb pattern formation by some mechanism that is not dependent on altered cell proliferation. In the *Bufo* system, the extent of limb growth was dependent on the concentration and duration of retinoid treatment. At increasing concentrations of retinol palmitate and duration of exposure, the level at which the limb was duplicated became more and more proximal, and finally, a complete limb grew from a distal amputation.[119] The concentration of retinol palmitate required to induce the growth of a complete limb decreased as the amputation became more proximal. There was a greater effect in the forelimb than in the hindlimb. Retinoic acid was the most active compound in inducing these effects in a series of retinoids tested. The conclusion was drawn from this study that pattern and growth may be specified by different mechanisms, for changes in pattern do not require cell division. Positional information may be encoded by the lengths of sugar chains on surface glycoproteins, therefore the effects of retinoids on glycoprotein synthesis are probably responsible for these phenomena.[123] Inhibition of oligosaccharide assembly by the antibiotic tunicamycin is followed by abnormal cell-cell adhesion and tissue organization, presumably the result of cell surface marker changes.[124]

There is also recent evidence that local application of retinoic acid to the limb bud mimics the action of the polarizing region.[125]

VIII. EFFECTS OF VITAMIN A ON GLYCOSYLATION

Glycosylation of lipids and proteins is necessary for cell-cell and cell-substratum contact and adhesion.[126] It is also involved in the processing and secretion of proteins and may play a role in the regulation of gene expression.[127]

Glycolipids and glycoproteins are involved in the expression of information on cell surfaces and glycoproteins are components of all mucous secretions.

Proteins are glycosylated in the ER, where this process is directed by the sequence and conformation of the polypeptide[128] before transfer to the Golgi and further modification. The process requires a lipophilic carrier molecule to transfer hydrophobic sugar-nucleotides into the lipid environment of the ER[129] and the two carrier molecule systems involved are those utilizing the polyisoprenoid derivatives dolichol mannosyl phosphate (Dol-P-Man)[129] and mannosyl retinol phosphate (MRP),[130-136] the latter compound being a metabolite of retinol. There is evidence that the core sequence of asparagine-linked oligosaccharides may be assembled in the same manner for all eukaryotic glycoproteins, but the processing and elongation of such sequences reveal species-related evolutionary differences.[123] This is not unexpected, for as well as glycoproteins playing an essential role in cell-cell adhesion and recognition, there is evidence that the length of the oligosaccharide chain encodes positional information in regenerating tissues.[119] In fact, a sulfated glycoprotein is important in the early stages of embryonic development.[137] Because many glycoproteins are sulfated, the process of glycoprotein production requires a source of active sulfate, 3'-phosphoadenosine-5'-phosphosulfate (PAPS).

In vitamin A deficiency states, there is decreased glycoprotein production[136] as manifested by decreases in glycoprotein secretions and serum α_1-macroglobulin glycosylation.[136,137] There is a decreased number of goblet cells in the rat intestine[138] and in ocular conjunctiva,[139] as well as evidence of altered cell adhesion.

The glycosylation defect in vitamin A deficiency states occurs at the level of transfer of sugars to newly synthesized proteins and lipids in the ER; there is an accumulation of probably immature small molecular weight oligosaccharide lipids in this condition.[140,141] There is no evidence that the subsequent modification of glycoproteins by glycosyltransferases in the Golgi prior to export is defective.[51] There is also an apparent deficiency of PAPS in vitamin A deficiency which leads to decreased synthesis of glycoproteins containing sulfur-linked sugars.[142]

It is therefore not surprising that in vitamin A deficiency states, immune functioning is defective[1] and the process of carcinogenesis is facilitated, for cells undergoing growth have increased levels of heparin sulfate and retinoids may modulate cell behavior by cell surface changes in heparin sulfate and other glycosaminoglycans (GAG).[144] Interestingly, the cell surface receptor for EGF is a glycoprotein.[90] GAG are increased in fetal liver; chondroitin sulfate and hyaluronic acid are increased in liver tumors while heparin sulfate is decreased. Heparin sulfate may regulate cell proliferation in healthy liver.[144] Furthermore, tumor-promoting agents such as phorbol esters and teleocidin oppose the effects of retinoids on the synthesis of specific glycoproteins involved in cell adhesion in cultured cells.[132]

IX. EFFECTS OF VITAMIN A ON THE IMMUNE SYSTEM

In recent years, evidence has accumulated that retinoids are essential for effective cell-mediated and humoral immunity. Thus, vitamin A is an immune adjuvant.[92]

The effect of retinoids on cell-mediated immunity are probably either due to stimulation of T cell proliferation,[92] for example an increased generation of killer T cells in allogeneic or syngeneic tumor systems,[148] or alternatively to elimination of a subset of suppressor T cells.[92] β-Carotene may also affect immune functioning, probably by conversion to retinol.[43]

The effects of retinoids on the immune system are probably mediated at least in part by the effects of retinoids on intracellular glycosylation reactions.[123] Immunoglobulins have asparagine-linked oligosaccharides. The core sequence of these may be assembled in the same manner for all eukaryotic glycoproteins, however, the processing and elongation of the sequences reveal evolutionary differences correlated with species-related biological functions. Macrophages have surface mannose receptors and mannosylated immunoglobulins thus opsonize pathogens that lack mannose-containing structures on their cell surfaces for macrophage attack. Once ingested, the mannose phosphate is a marker that segregates proteins in lysosomes for destruction.

Some of the antitumor activity of retinoids is doubtlessly mediated by their effects on the immune system, however this sometimes results in enhancement of tumor growth.[147,148] The growth and development of a transplantable murine melanoma was inhibited by oral or intraperitoneal vitamin A,[149] while the effect was not seen with methylcholanthrene (MC)-induced murine tumors, which are less strongly immunogenic.[149] Retinoids also decrease the amount of radiation required to control murine tumor growth and can interact with antilymphocyte serum in its effects on tumor growth.[149] Combined treatment with retinol palmitate and 5-fluorouracil (FUra) decreased tumor growth and increased survival time in ICR/JCL mice given S/C allotransplantable sarcoma 180 cells.[148] Implantation of fibrosarcomas was also inhibited by retinyl palmitate, especially by combined treatment with FUra.

Possibly this combined therapy, which may involve effects on the immune response, may prove important in the prevention of tumor recurrences and metastases.

A further effect of retinoids relevant to the control of neoplasia in man is the reversal of postoperative immunosuppression.[94] It is therefore essential that the vitamin A status of the patient undergoing surgery for neoplasia is adequate.

For a lucid, current review of the immunological mechanisms of disease, the reader is referred to Taussig.[145] A more detailed discussion of the role of the immune system in neoplasia is given by Pitot.[146]

X. VITAMIN A AND CARCINOGENESIS

There is now general agreement that carcinogenesis in most tissues and species is a multistep process.[150] In mouse skin, the stages of initiation and promotion were described and there is now evidence that promotion itself can be subdivided into two stages.[116,151] Initiation and promotion have likewise been demonstrated in tissue culture transformation systems[152] and in carcinogenesis of various animal tissues in addition to skin.[150] Initiation is rapid and irreversible and is believed to be a consequence of mutational changes in DNA induced by the interactions of highly reactive electrophilic carcinogen metabolites with nucleophilic regions of nucleic acids.[153,154]

The anticarcinogenic effects of retinoids could involve this "initiation" step either by interfering with the production of the electrophilic carcinogenic metabolites or by balancing alternative metabolic pathways leading to unreactive derivatives that are readily excreted. Retinoids could also react themselves with the electrophilic species generated. The influence of vitamin A status on the metabolism of xenobiotics is therefore of some interest; the effect in any instance will obviously depend upon the particular balance of activation/ inactivation steps. Vitamin A is a lipophilic compound which interacts with cellular membranes[18] and would therefore be expected to influence metabolism by the mixed function oxidases that metabolize xenobiotics. These are found to be intimately associated with the lipid membranes of the SER. In fact vitamin A increases the activity of an isomer of isocitric dehydrogenase associated with mitochondrial membranes in a dose-dependent fashion and is believed to be a physiological regulator of this enzyme.[155] In vitamin A deficiency states, there are reports of decreases in liver weight, microsomal protein, and cytochrome P_{450}.[156-158] There are also reports of decreases in demethylase activity[156] and increases in hexobarbital sleeping time,[158] all suggesting a decrease in biotransformation reactions. Phase II conjugation reactions were inhibited to a greater extent than Phase I oxidative reactions[157] and there is evidence that production of PAPS is decreased in vitamin A deficiency,[142] even though sulfation of mucopolysaccharides was not significantly affected by vitamin A status.[142] In contrast, there are reports of vitamin A decreasing mixed function oxidase activity.[159,160] Possibly, a certain level of lipophilic retinoids is necessary for the integrity of the membrane-associated electron transport chains that are involved in microsomal oxidative metabolism, but an excess causes perturbation of the process. Thus there is evidence that in vitro addition of retinol interferes with oxidative metabolism.[156]

An interesting report by Dickins and Sorof[161] of the effect of the retinoid retinylidene dimedone on the transformation of cultured rat mammary gland tissue revealed that this retinoid inhibited transformation by procarcinogens, suggesting that there was inhibition of carcinogen metabolism, but failed to inhibit transformation by directly applied putative active carcinogen metabolites if administered concurrently. The conclusion was drawn that the major inhibiting action of the retinoids was on promotion, although retinoids have no influence on tissue uptake or metabolism of the potent promoter TPA,[131] which is not believed to require metabolic activation to exert tumor-promoting activity. Further discussion of interactions between retinoids and tumor promoters is to be found in the section on the effects of vitamin A on cell division and retinol has recently been shown to reduce AFB_1-, and cyclophosphamide-induced sister chromatid exchanges and cell cycle delays in Chinese hamster V79 cells in a dose- and time-dependent manner,[162] although, by itself, it had no effect on either parameter even in the presence of S9 mix.[162]

The effect of retinol on the mutagenicity of 2-aminofluorene (2AF) and 2-acetylamino-fluorene (2-AAF) in the Salmonella/microsome assay has recently been investigated.[163,164] Low doses of retinol (2 to 20 μg per plate) significantly increased the mutagenicity of both compounds while larger doses had no effect or led to a gradual decrease in the mutagenicity of 2AF. Retinol was not mutagenic itself and did not affect bacterial survival.[163] The authors attributed the effect to an altered pattern of metabolism of 2AF and 2AAF in the presence of retinol for several forms of cytochrome P_{450} are involved in the transformation of 2AAF.[163] The effect is obviously not due to increased permeability of the bacteria to the mutagens as it was not observed at higher concentrations of retinol.[163]

A further means by which retinoids could affect initiation of carcinogenesis is by their effects on the metabolism of vitamin C[165] and vitamin D.[166] In fact, there is evidence that vitamin C prevents nitrosamine carcinogenesis by preventing the generation of nitrosamines in the stomach by the reaction of amines and nitrite.[167] As vitamin A deficiency decreases ascorbate synthesis in the rat,[165] it therefore could enhance nitrosamine carcinogenesis in this species by this mechanism.

XI. INHIBITION OF CARCINOGENESIS BY RETINOIDS

Since the initial finding that rats fed a diet deficient in vitamin A had an increased incidence of stomach tumors,[1] there have been numerous reports of retinoids inhibiting carcinogenesis in experimental animals. The inhibition is effective against carcinogenesis induced by polycyclic aromatic hydrocarbons (PAH),[1,161,168-170] aromatic amines,[163,164] nitrosamines,[82,86,171,172] azo dyes,[173] and miscellaneous other chemical carcinogens[1,3,11-17] in various target tissues such as skin,[168,174,175] breast,[66,86,161,169,176-182] colon,[82,172] bladder,[183,184] and lung.[170,185,186] There is a burgeoning literature on the topic.[1,3,11-17,187] Retinoids prevent transformation of cells in culture by carcinogens[98,120,121,161,186,188,189] and promoting agents.[110,131,174,182,188,196] There are reports of inhibition of radiation- and virally induced tumors;[190] retinoids also inhibit transformation of cells by sarcoma growth factor.[111]

In recent years there have also been numerous epidemiological studies[191,192] on human populations linking vitamin A status and cancer incidence; earlier reports have been reviewed by Peto et al.,[43] but there have been several more published since then.[191,192]

The possibility that retinoids could be used for the chemoprophylaxis of various neoplasms in susceptible populations has received serious attention recently,[43] and to this end there has been much work done on the development of synthetic, nontoxic retinoids that would be suitable for long-term administration to human populations.[12,187]

A large number of recent experimental studies have been concerned with the effects of retinoids on the induction of breast tumors. Dickins and Sorof[164] studied the effects of retinylidene dimedone on the transformation of organ cultures of mammary glands from BALB/c mice. In this system, transformation was not prevented when the retinoid was administered either before or concurrently with the procarcinogens benzo[a]pyrene (BAP) and 2-AAF. However, if retinoids were administered subsequently, transformation was inhibited. If the dose of the procarcinogens was increased, the inhibition by the retinoid was overcome. Transformations by activated carcinogens such as the diol-epoxide of BAP and acetoxy-AAF were not inhibited.[164] Thus the inhibition was only effective in the promotional or facilitational phases of carcinogenesis against the typically low levels of activated carcinogens generated endogenerously in the target cells. This finding suggests that retinoids may find application in the chemoprevention of "spontaneous" breast tumors in women, where the unknown carcinogens are presumably present at low doses and where the latent period for tumor induction is correspondingly long.

To this end, McCormick and Moon[184] have recently studied the influence of delayed administration of retinyl acetate on mammary carcinogenesis in rats induced by MNU. Retinyl acetate was administered in the diet commencing 1, 4, or 8 weeks after a low (25 mg) or high (50 mg) dose MNU given intravenously. At the high dose of MNU, the retinyl acetate was most effective if treatment began 1 week following carcinogen exposure; there was no effect of inhibition of carcinogenesis if it was delayed for 8 weeks after carcinogen dosing. However, with the low dose of carcinogen, an 8-week delay before retinyl acetate treatment was initiated still resulted in effective chemoprevention.[184] As tumor latency is related to carcinogen dose,[150] the period of delay permissable before therapy is administered is a function of tumor latency, a finding that suggests that retinoid treatment of populations at risk of developing breast cancer due to probably low doses of unknown environmental carcinogens may be effective in preventing tumors.

Inhibition of carcinogenesis by retinoids in this system is correlated with the presence of cRABP in the target cells.[86] Interestingly, cRABP is increased by estradiol administration.[84] Such a mechanism may be responsible for the protective effect of the Pill against breast pathology in human populations.[193] A further study showed that inhibition of dimethylbenzanthracene (DMBA)-induced rat mammary tumors by retinyl acetate was equally effective in intact and ovariectomized rats.[179] Thus retinyl acetate inhibits both ovarian hormone-responsive and nonresponsive tumors. However, a more recent study by this group reveals enhanced inhibition of mammary carcinogenesis by retinoid treatment and ovariectomy.[180]

However, a note of caution should be sounded, for retinyl acetate could either stimulate or inhibit mammary tumor cell growth in vitro, depending on the concentration present.[7,181] There is thus a possibility that tumor development may be enhanced rather than inhibited by retinoid therapy. Retinyl acetate and selenium have recently been shown to exert a combined effect in the chemoprevention of mammary tumors.[176]

Retinyl methyl ether added to the culture medium prevents epithelial changes induced in organ cultures of hamster trachea by crocidolite and amosite asbestos,[186] a finding that suggests that prospective intervention studies with retinoids in asbestos-exposed workers may be effective in decreasing their risk of lung cancer. A recent epidemiological study in which retrospective dietary and smoking data were collected from lung cancer patients and controls indicated that vitamin A intake is associated with a lower relative risk of lung cancer in heavy smokers.[185] Bronchial carcinoma patients have decreased plasma vitamin A,[191,194] especially if the tumors were of the oat cell or squamous cell varieties.[191,194] However, this finding should be interpreted with caution as the low plasma vitamin A may be a consequence, rather than a cause of, the tumor. Nevertheless, Wald et al.[192] recently showed in a prospective study that serum vitamin A levels were lower in patients subsequently developing cancer rather than in controls. The effect was independent of age, smoking status, or serum cholesterol and was greatest for lung cancer and cancer of the GI tract.

Skin tumor induction in mice by DMBA or UVB is inhibited by carotenoid administration,[168] and tumor induction in mouse skin by "initiation" with DMBA and "promotion" with TPA is inhibited by retinoic acid or the synthetic retinoid Ro 10-9359.[175] Interestingly, skin tumor inhibition was also produced by phytoene and canthaxanthine,[168] which are two carotenoids that do not possess vitamin A activity. However, this effect was only seen with skin tumors induced by UVB (290 to 320 nm) and these carotenoids can quench singlet-excited oxygen. These two carotenoids did not inhibit DMBA-induced skin tumors.[168]

Retinoic acid[174] and Ro 10-9359[175] can inhibit carcinogenesis in mouse skin in the DMBA/ TPA initiation/promotion system. Both compounds are more effective if given in combination with steroids,[174,175] which are inhibitors of tumor promotion.[195] While dexamethasone decreased TPA-induced DNA synthesis, retinoic acid did not affect DNA synthesis,[174] suggesting that retinoids and steroids both act antipromotionally, but by different mechanisms. However, inhibition of promotion by both compounds is reversible.

There are other recent reports of retinoid inhibition of various experimental tumors in the liver, colon, and bladder. Thus, Daoud and Griffin[173] showed that dietary retinoic acid inhibited 3'methyl-4-dimethylaminoazobenzene-induced liver tumors in rats. Retinyl acetate also decreased the incidence of spontaneous hepatomas in mice.[182] Hepatocarcinogens[196] and alcohol[40] decrease the vitamin A content of the liver. Alcoholics and others at risk of developing hepatocarcinoma may decrease this risk if they ensure that their vitamin A intake is adequate.

Several retinoids (N-ethylretinamide, N-2 hydroxyethyl retinamide, N-[4-hydroxyphenyl]-all-trans-retinamide or retinyl acetate) inhibited the induction of colon cancer in F344 rats by MNU.[172] This finding agrees with other findings of Silverman et al.[82] and suggests that the local concentration of retinoid at the site of tumor induction must be high.

The sex-adjusted relative risk for the development of bladder cancer is increased for humans with lower levels of an index of vitamin A intake,[183] and dibenzylnitrosamine (DBA)-induced squamous cell carcinomas of the bladder are inhibited by retinoids.[171] A vitamin A-deficient diet containing lactose, which is calculogenic, is also highly effective in inducing bladder tumors in rats.[184] These findings, taken together with epidemiological and experimental findings on the induction of lung cancer, suggest that smokers, who are at risk of developing both lung and bladder cancer, are a group that must ensure that they have an adequate vitamin A intake.

There are many reports of retinoids opposing the actions of chemical carcinogens, TPA, radiation, and even viruses on the transformation of various cell lines in culture. Transformation in JB-6 cells in culture is accompanied by loss of density-dependent inhibition of cell growth and anchorage-independent growth. These properties have been associated with the loss of a 180K cell surface glycoprotein.[131] Loss of this glycoprotein is enhanced by treatment with the tumor-promoting agent TPA. This effect is opposed by retinoids, which affect glycoprotein synthesis.[1,131]

Transformation of tissue-cultured fibroblasts by sarcoma growth factor is also opposed by retinoids.[111] Retinoids likewise block ornithine decarboxylase induction in cells treated with TPA, sarcoma growth factor, or EGF.[110] There is recent evidence that sarcoma growth factor activates cellular protein kinases which phosphorylate tyrosine residues in cellular proteins, thus modulating their activity.[105] One protein thus affected is the Na^+/K^+-activated ATPase of the cell membrane; it is interesting that retinoids modulate the activity of this enzyme.[197]

The effects of retinoic acid in cells in tissue culture may be mediated by the metabolite 5,6-epoxyretinoic acid, for the epoxide has an increased ability to oppose TPA-induced effects.[188]

XII. ENHANCEMENT OF CARCINOGENESIS BY VITAMIN A

While most reports concerning the effects of retinoids on carcinogenesis refer to inhibition of tumor induction, there are sporadic instances, sometimes in the same reports,[82,198] of tumor enhancement. Thus pancreatic tumors induced in hamsters by N-nitroso-bis (2-oxypropyl) amine (BOP) may be either inhibited or enhanced by retinoid treatment.[198] The effect was sex dependent, the retinoids being inhibitory in females, but not in males.[198] Likewise, colonic tumors induced in F344 rats by either 1,2-dimethylhydrazine or MNU were either inhibited or enhanced by retinoids.[82] The effect was seen especially if the retinoid treatment followed carcinogen administration in this instance. Locally administered retinoids were more effective in either inhibition or enhancement. Other experimental systems in which retinoids enhance tumor induction are the induction by DMBA of tumors on rabbit ears,[93] the induction by 3MC of lymphoid tumors in PBA/2 mice,[148] the induction of skin tumors by DMBA or UV light,[199] the induction of breast tumors,[200] and DMBA-induced tumors in the hamster cheek pouch.[93]

A semisynthetic compound, phorbol-12-retinoate-13-acetate has been produced, which combines the antipromoting principle of vitamin A with a tumor-promoting phorbol ester.[116] PRA is a nonpromoting irritant mitogen which induces epidermal proliferation by a mechanism involving increased synthesis of prostaglandin E. Using this compound, evidence has been obtained that tumor promotion in skin can be divided into two stages, with the mitogenic PRA being an effective second-stage promoter.[116]

The mechanism by which retinoids enhance tumor development is a matter for conjecture. They may accelerate procarcinogen metabolism to electrophiles or inhibit deactivating pathways catalyzed by mixed function oxidases, they may affect immune functioning in such a way as to stimulate tumor growth, they may increase carcinogen access to tissues because of their membrane effects, or they may act as mitogens and accelerate promotion. There is also a report that *all-trans* retinoic acid enhances 3MC-induced lymphoid cell transformation by inhibiting the antitumor effect of interferon.[148]

REFERENCES

1. **Mayer, H., Bollag, W., Hänni, R., and Rüegg, R.,** Retinoids, a new class of compounds with prophylactic and therapeutic activities in oncology and dermatology, *Experientia,* 24, 1105, 1978.
2. **McGilvery, R. W.,** *Biochemistry: A Functional Approach,* W. B. Saunders, Philadelphia, 1970, 639.
3. **Sporn, M. B., Newton, D. L., Smith, J. M., Acton, N., Jacobsen, A. E., and Brossi, A.,** Retinoids and cancer prevention: the importance of the terminal group of the retinoid molecule in modifying activity and toxicity, in *Carcinogens: Identification and Mechanisms of Action,* Griffin, A. C. and Shaw, C. R., Eds., Raven Press, New York, 1979, 441.
4. **Goodman, DeW. S.,** Vitamin A and retinoids: recent advances. Introduction, background, and general overview, *Fed. Proc. Fed. Am. Soc. Exp. Biol.,* 38, 2501, 1979.
5. World Health Organization, Control of vitamin A deficiency and xeropthalmia, WHO Tech. Rep. Ser. No. 672, Geneva, 1982.
6. **Epstein, S. S. and Swartz, J. B.,** Fallacies of lifestyle cancer theories, *Nature (London),* 289, 127, 1981.
7. **Levy, R. I. and Moskowitz, J.,** Cardiovascular research: decades of progress, a decade of promise, *Science,* 217, 121, 1982.
8. **Peto, R.,** Distorting the epidemiology of cancer: the need for a more balanced overview, *Nature (London),* 284, 297, 1980.
9. Editorial, Hazard export: a case for international concern, *Nature (London),* 273, 415, 1978.
10. **Maugh, T. H., II,** Vitamin A: potential protection from carcinogens, *Science,* 186, 1198, 1974.
11. **Sporn, M. B.,** Retinoids: new developments in their mechanism of action as related to control of proliferative diseases, in *Retinoids: Advances in Basic Research and Therapy,* Orfanos, C. E., Braun-Falco, O., Farber, E. M., Grupper, C. H., Polano, M. K., and Schuppli, R., Eds., Springer Verlag, Basel, 1981, 73.
12. **Sporn, M. B.,** *Pharmacological prevention of carcinogenesis by retinoids, in Carcinogenesis,* Vol. 2, *Mechanisms of Tumor Promotion and Cocarcinogenesis,* Slaga, T. J., Sivak, A., and Boutwell, R. K., Eds., Raven Press, New York, 1978, 545.
13. **Sporn, M. B. and Newton, D. L.,** Chemoprevention of cancer with retinoids, *Fed. Proc. Fed. Am. Soc. Exp. Biol.,* 38, 3528, 1979.
14. **Sporn, M. B., Dunlop, N. M., Newton, D. L., and Smith, J. M.,** Prevention of chemical carcinogenesis by vitamin A and its synthetic analogs (retinoids), *Fed. Proc. Fed. Am. Soc. Exp. Biol.,* 35, 1332, 1976.
15. **Sporn, M. B.,** Retinoids and carcinogenesis, *Nutr. Rev.,* 35, 65, 1977.
16. **Sporn, M. B.,** Vitamin A and its analogs (retinoids) in cancer prevention, in *Nutrition and Cancer,* Vol. 6, *Current Concepts in Nutrition,* Winick, M., Ed., John Wiley & Sons, New York, 1977, 119.
17. **Sporn, M. B.,** Retinoids and cancer prevention, in *Carcinogenesis,* Vol. 5, *Modifiers of Chemical Carcinogenesis,* Slaga, T. J., Ed., Raven Press, New York, 1980, 99.
18. **Dingle, J. T. and Lucy, J. A.,** Vitamin A, carotenoids and cell functions, *Biol. Rev.,* 40, 422, 1965.
19. **DeLuca, H. F.,** Retinoic acid metabolism, *Fed. Proc. Fed. Am. Soc. Exp. Biol.,* 38, 2519, 1979.

20. **Takruri, H. R. H. and Thurnham, D. I.**, Effect of zinc deficiency on the conversion of β-carotene to retinol as indicated by liver stores, *Proc. Nutr. Soc.*, 41, 53A, 1982.

21. **Ross, A. C.**, Retinol esterification by rat liver microsomes. Evidence for a fatty acyl coenzyme A: retinolacyltransferase, *J. Biol. Chem.*, 257, 2453, 1982.

22. **Sklan, D., Blaner, W. S., Adachi, N., Smith, J. E., and Goodman, DeW. S.**, Association of cellular retinol-binding protein and several lipid hydrolase activities with a vitamin A-containing high-molecular weight lipid-protein aggregate from rat liver cytosol, *Arch. Biochem. Biophys.*, 214, 35, 1982.

23. **Vahlqvist, A., Berne, B., and Berne, C.**, Skin content and plasma transport of vitamin A and β-carotene in chronic renal failure, *Eur. J. Clin. Invest.*, 12, 63, 1982.

24. **Bernard, A. M., Moreau, D., and Lauwerys, R. R.**, Latex immunoassay of retinol-binding protein, *Clin. Chem.*, 28, 1167, 1982.

25. **Ott, D. B. and Lachance, P. A.**, Retinoic acid — a review, *Am. J. Clin. Nutr.*, 32, 2522, 1979.

26. **Vane, F. M. and Bugge, C. J. L.**, Identification of 4-oxo-13-*cis* retinoic acid as the major metabolite of 13-*cis*-retinoic acid in human blood, *Drug Metab. Dispos.*, 9, 515, 1981.

27. **Sato, M. and Lieber, C. S.**, Increased metabolism of retinoic acid after chronic ethanol consumption in rat liver microsomes, *Arch. Biochem. Biophys.*, 213, 557, 1982.

28. **Sietsma, W. K. and DeLuca, H. F.**, Retinoic acid 5,6-epoxidase. Properties and biological significance, *J. Biol. Chem.*, 257, 4265, 1982.

29. **Roberts, A. B. and Frolik, C. A.**, Recent advances in the *in vivo* and *in vitro* metabolism of retinoic acid, *Fed. Proc. Fed. Am. Soc. Exp. Biol.*, 38, 2524, 1979.

30. **Napoli, J. L. and McCormick, A. M.**, Tissue dependence of retinoic acid metabolism *in vivo*, *Biochim. Biophys. Acta*, 666, 167, 1981.

31. **Roberts, A. B., Nichols, M. D., Newton, D. L., and Sporn, M. B.**, *In vitro* metabolism of retinoic acid in hamster intestine and liver, *J. Biol. Chem*, 254, 6296, 1979.

32. **Roberts, A. B., Frolik, C. A., Nichols, M. D., and Sporn, M. B.**, Retinoid-dependent induction of the *in vivo* and *in vitro* metabolism of retinoic acid in tissues of the vitamin A-deficient hamster, *J. Biol. Chem.*, 254, 6303, 1979.

33. **Zile, M. H., Inhorn, R. C., and DeLuca, H. F.**, Metabolism *in vivo* of all *trans*- retinoic acid. Biosynthesis of 13-*cis*-retinoic acid and all-*trans* and 13-*cis*-retinylglucuronides in the intestinal mucosa of the rat, *J. Biol. Chem.*, 257, 3544, 1982.

34. **Napoli, J. L., Khalil, J., and McCormick, A. M.**, Metabolism of 5,6-epoxyretinoic acid *in vivo:* isolation of a major intestinal metabolite, *Biochemistry*, 21, 1942, 1982.

35. **Zile, M. H., Inhorn, R. C., and DeLuca, H. F.**, Metabolites of all-*trans*-retinoic acid in bile: identification of all-*trans* and 13-*cis*-retinoylglucuronides, *J. Biol. Chem.*, 257, 3537, 1982.

36. **Vahlqvist, A.**, Retinoids in human epidermis, *Ann. N.Y. Acad. Sci.*, 359, 366, 1981.

37. **Huque, T.**, Excretion of radioactive metabolites of retinol as an index of vitamin A status in rats, *Nutr. Rep. Int.*, 24, 171, 1981.

38. **Rietz, P., Wiss, O., and Weber, F.**, Metabolism of vitamin A and the determination of vitamin A status, *Vitam. Horm.*, 32, 237, 1974.

39. **Nakano, K. and Morita, A.**, Redistribution of vitamin A in tissues of rats with imposed chronic confinement stress, *Br. J. Nutr.*, 47, 645, 1982.

40. **Leo, M. A. and Lieber, C. S.**, Hepatic vitamin A depletion in alcoholic liver injury, *N. Engl. J. Med*, 307, 597, 1982.

41. **McKenna, M. C. and Bieri, J. G.**, Tissue storage of vitamins A and E in rats drinking or infused with total parenteral nutrition solutions, *Am. J. Clin. Nutr.*, 35, 1010, 1982.

42. **Goodman, DeW. S.**, Vitamin A metabolism, *Fed. Proc. Fed. Am. Soc. Exp. Biol.*, 39, 2716, 1980.

43. **Peto, R., Doll, R., Buckley, J. D., and Sporn, M. B.**, Can dietary carotene materially reduce human cancer rates? *Nature (London)*, 290, 201, 1981.

44. **Bashor, M. M., Toft, D. O., and Chytil, F.**, *In vitro* binding of retinol to rat-tissue components, *Proc. Nat. Acad. Sci. U.S.A.*, 70, 3483, 1973.

45. **Smith, J. E., and Goodman, DeW. S.**, Retinol-binding protein and the regulation of vitamin A transport, *Fed. Proc. Fed. Am. Soc. Exp. Biol.*, 38, 2504, 1979.

46. **Pitt, G. A. J.**, The assessment of vitamin A status, *Proc. Nutr. Soc.*, 40, 173, 1981.

47. **Solomons, N. W. and Russell, R. M.**, The interaction of vitamin A and zinc: implications for human nutrition, *Am. J. Clin. Nutr.*, 33, 2031, 1980.

48. **Mathias, P. M.**, The effect of zinc supplementation on plasma levels of vitamin A and retinol-binding protein (RBP) in children recovering from protein-energy malnutrition (PEM), *Proc. Nutr. Soc.*, 41, 52A, 1982.

49. **Goodman, DeW. S.**, Vitamin A transport and delivery and the mechanism of vitamin A toxicity, in *Retinoids: Advances in Basic Research and Therapy*, Orfanos, C. E., Braun-Falco, O., Farber, E. M., Grupper, C. H., Polano, M. K., and Schuppli, R., Eds., Springer-Verlag, Basel, 1981, 31.

50. **Supopark, W. and Olson, J. A.**, Effect of ovral, a combination type oral contraceptive agent, on vitamin A metabolism in rats, *Int. J. Vitam. Nutr. Res.*, 45, 113, 1975.

51. **Morita, A. and Nakano, K.**, The effect of chronic immobilization stress on tissue distribution of vitamin A in rats fed a diet with adequate vitamin A, *J. Nutr.*, 112, 789, 1982.

52. **Harrison, E. H., Smith, J. E., and Goodman, DeW. S.**, Effects of vitamin A deficiency on the levels and distribution of retinol-binding protein and marker enzymes in homogenates and Golgi-rich fractions of rat liver, *Biochim. Biophys. Acta*, 628, 489, 1980.

53. **Joshi, H. C. and Misra, U. K.**, Effect of vitamin A on microtubules of rat brain and liver, *Int. J. Vitam. Nutr. Res.*, 51, 93, 1981.

54. **Morley, J. E., Damassa, D. A., Gordon, J., Pekary, A. E., and Hershmann, J. M.**, Thyroid function and vitamin A deficiency, *Life Sci.*, 22, 1901, 1978.

55. **Clarke, G. D.**, Vitamin A and cancer, *Lancet*, 1, 830, 1980.

56. **Mallia, A. K., Smith, J. E., and Goodman, DeW. S.**, Metabolism of retinol-binding protein and vitamin A during hypervitaminosis A in the rat, *J. Lipid Res.*, 16, 180, 1975.

57. **Kark, J. D., Smith, A. H., and Hames, C. G.**, Serum retinol and the inverse relationship between serum cholesterol and cancer, *Br. Med. J.*, 284, 152, 1982.

58. **Gerber, L. E. and Erdman, J. W., Jr.**, Retinoic acid and hypertriglyceridemia, *Ann. N. Y. Acad. Sci.*, 359, 391, 1981.

59. **Adachi, N., Smith, J. E., Sklan, D., and Goodman, DeW. S.**, Radioimmunoassay studies of the tissue distribution and subcellular localization of cellular retinol-binding protein in rats, *J. Biol. Chem.*, 256, 9471, 1981.

60. **Gawinowicz, M. A. and Goodman, DeW. S.**, Retinoid affinity label for the binding site of retinol-binding protein, *Biochemistry*, 21, 1899, 1982.

61. Editorial, Vitamin A and cancer, *Lancet*, 1, 575, 1980.

62. **Walker, A., Zimmerman, M. R., and Leakey, R. E. F.**, A possible case of hypervitaminosis A in *Homo erectus, Nature (London)*, 296, 248, 1982.

63. **Meeks, R. G., Zaharevitz, D., and Chan, R. F.**, Membrane effects of retinoids: possible correlation with toxicity, *Arch. Biochem. Biophys.*, 207, 141, 1981.

64. **Fell, H. B. and Dingle, J. T.**, Studies on the mode of action of excess vitamin A. Lysosomal protease and the degradation of the cartilage matrix, *Biochem. J.*, 87, 403, 1963.

65. **Bryden, W. L., Cumming, R. B., and Balnave, D.**, Influence of vitamin A status on response of chickens to aflatoxin B₁ and changes in liver lipid metabolism associated with aflatoxins, *Br. J. Nutr.*, 41, 529, 1979.

66. **Maiorana, A. and Gullino, P. M.**, Effect of retinyl acetate on the incidence of mammary carcinomas and hepatomas in mice, *J. Natl. Cancer Inst.*, 64, 655, 1980.

67. **Macapinlac, M. P. and Olson, J. A.**, A lethal hypervitaminosis A syndrome in young monkeys *(Macacus fascicularis)* following a single intramuscular dose of a water-miscible preparation containing vitamins A, D₂ and E, *Int. J. Vitam. Nutr. Res.*, 51, 331, 1981.

68. Anon., The function of retinol and retinoic acid in the testes, *Nutr. Rev.*, 40, 187, 1982.

69. **Sani, B. P. and Hill, D. L.**, A retinoic acid-binding protein from chick embryo skin, *Cancer Res.*, 36, 409, 1976.

70. **Muto, Y., Moriwaki, H., and Omori, M.**, *In vitro* binding affinity of novel synthetic polyprenoids (polyprenoic acids) to cellular retinoid-binding proteins, *Gann*, 72, 974, 1981.

71. **Takase, S., Ong, D. E., and Chytil, F.**, Cellular retinol-binding protein allows specific interaction of retinol with the nucleus *in vitro, Proc. Natl. Acad. Sci. U.S.A.*, 76, 2204, 1979.

72. **Fex, G. and Johannesson, G.**, Purification and partial characterization of a cellular retinol-binding protein from human liver, *Biochim. Biophys. Acta*, 714, 536, 1982.

73. **Ong, D. E. and Chytil, F.**, Immunochemical studies on cellular vitamin A-binding proteins, *Ann. N.Y. Acad. Sci.*, 359, 415, 1981.

74. **Chytil, F. and Ong, D. E.**, Mediation of retinoic acid-induced growth and anti-tumour activity, *Nature (London)*, 260, 49, 1976.

75. **Banerjee, C. K. and Sani, B. P.**, Cellular uptake of retinoic acid *in vitro, Biochem. J.*, 190, 839, 1980.

76. **Sani, B. P. and Banerjee, C. K.**, Cellular and subcellular uptake of retinoic acid and its mediation by retinoic acid binding protein, *Ann. N.Y. Acad. Sci.*, 359, 420, 1981.

77. **Gmeiner, B. and Zerlauth, G.**, Enhanced binding of retinoic acid to its cellular binding protein in Lewis lung tumor cytosol by alkaline phosphatase treatment, *Hoppe-Seyler's Z. Physiol. Chem.*, 363, 337, 1982.

78. **Sani, B. P. and Donovan, M. K.**, Localization of retinoic acid-binding protein in nuclei and the nuclear uptake of retinoic acid, *Cancer Res.*, 39, 2492, 1979.

79. **Liau, G., Ong, D. E., and Chytil, F.**, Interaction of the retinol/cellular retinol-binding protein complex with isolated nuclei and nuclear components, *J. Cell Biol.*, 91, 63, 1981.

80. **Rask, L., Anundi, H., Böhme, J., Eriksson, U., Ronne, J., Sego, K., and Peterson, P.**, Structural and functional studies of vitamin A-binding proteins, *Ann. N.Y. Acad. Sci.*, 359, 79, 1981.

81. **Ong, D. E. and Chytil, F.**, Changes in levels of cellular retinol- and retinoic acid binding proteins of liver and lung during perinatal development of rat, *Proc. Natl. Acad. Sci. U.S.A.*, 73, 3976, 1976.

82. **Silverman, T., Katayama, S., Zelenakas, K., Lauber, J., Musser, T. K., Reddy, M., Levenstein, M. J., and Weisburger, J. H.**, Effect of retinoids on the induction of colon cancer in F344 rats by N-methyl-N-nitrosourea or by 1,2-dimethylhydrazine, *Carcinogenesis*, 2, 1167, 1981.

83. **Muto, Y. and Omori, M.**, A novel cellular retinoid-binding protein, F-type, in hepatocellular carcinoma, *Ann. N.Y. Acad. Sci.*, 359, 91, 1981.

84. **Mehta, R. G. and Moon, R. C.**, Hormonal regulation of retinoic acid-binding proteins in the mammary gland, *Biochem. J.*, 200, 591, 1981.

85. **Lotan, R., Ong, D. E., and Chytil, F.**, Comparison of the level of cellular retinoid-binding proteins and susceptibility to retinoid-induced growth inhibition of various neoplastic cell lines, *J. Natl. Cancer Inst.*, 64, 1259, 1980.

86. **Mehta, R. J., McCormick, D. L., Carny, W. L., and Moon, R. C.**, Correlation between retinoid inhibition of N-methyl-N-nitrosourea-induced mammary carcinogenesis and levels of retinoic acid binding proteins, *Carcinogenesis*, 3, 89, 1982.

87. **Bollag, W. and Matter, A.**, From vitamin A to retinoids in experimental and clinical oncology, *Ann. N.Y. Acad. Sci.*, 359, 9, 1981.

88. **Schroder, E. W., Rapaport, E., Kabcenell, A. K., and Black, P. H.**, Growth inhibitory and stimulatory effects of retinoic acid on murine 3T3 cells, *Proc. Natl. Acad. Sci. U.S.A.*, 79, 1549, 1982.

89. **Marchok, A. C., Clark, J. N., and Klein-Szanto, A.**, Modulation of growth, differentiation, and mucous glycoprotein synthesis by retinyl acetate in cloned carcinoma cell lines, *J. Natl. Cancer Inst.*, 66, 1165, 1981.

90. **Kaneko, Y.**, Antagonistic action of retinoic acid and teleocidin on the proliferation and epidermal growth factor binding of rat hepatoma cells, *Int. J. Cancer*, 27, 841, 1981.

91. **Kinzel, V., Krombholz-Nolinski, I., Stöhr, M., and Richards, J.**, Effects of retinoic acid and tumor promoter 12-0-tetradecanoylphorbol-13-acetate (TPA) on the cell cycle of HeLa cells, *Int. J. Cancer*, 29, 93, 1982.

92. **Athanassiades, T. J.**, Adjuvant effect of vitamin A palmitate and analogs on cell-mediated immunity, *J. Natl. Cancer Inst.*, 67, 1153, 1981.

93. **Bollag, W.**, Vitamin A and vitamin A acid in the prophylaxis and therapy of epithelial tumours, *Int. J. Vitam. Res.*, 40, 299, 1970.

94. **Cohen, B. E., Gill, G., Cullen, P. R., and Morris, P. J.**, Reversal of postoperative immunosuppression in man by vitamin A, *Surg. Gynecol. Obstet.*, 149, 658, 1979.

95. **Lasnitzki, I.**, The influence of hypervitaminosis on the effect of 20-methyl-cholanthrene on mouse prostate glands grown *in vitro*, *Br. J. Cancer*, 9, 434, 1955.

96. **Dover, D. and Koeffler, H. P.**, Retinoic acid enhances growth of human early erythroid progenitor cells *in vitro*, *J. Clin. Invest.*, 69, 1039, 1982.

97. **Dover, D. and Koeffler, H. P.**, Retinoic acid enhances colony-stimulating factor-induced clonal growth of normal human myeloid progenitor cells *in vitro*, *Exp. Cell Res.*, 138, 193, 1982.

98. **Lotan, R., Neumann, G., and Lotan, D.**, Characterization of retinoic acid-induced alterations in the proliferation and differentiation of a murine and a human melanoma cell line in culture, *Ann. N.Y. Acad. Sci.*, 359, 150, 1981.

99. **Kubilus, J., Rand, R., and Baden, H. P.**, Effects of retinoic acid and other retinoids on the growth and differentiation of 3T3-supported human keratinocytes, *In Vitro*, 17, 786, 1981.

100. **Zil, S.**, Vitamin A acid effects on epidermal mitotic activity, thickness and cellularity in the hairless mouse, *J. Invest. Dermatol.*, 59, 228, 1972.

101. **Elias, P. M., Fritsch, P. O., Lampe, M., Milliams, M. L., Brown, B. E., Lemaric, M., and Grayson, S.**, Retinoid effects on epidermal structure, differentiation and permeability, *Lab. Invest.*, 44, 531, 1981.

102. **Otto, A. M., Ulrich, M.-O., and DeAsua, L. J.**, Epidermal growth factor initiates DNA synthesis after a time-dependent sequence of regulatory events in Swiss 3T3 cells - interactions with hormones and growth factors, *J. Cell Physiol.*, 108, 145, 1981.

103. **Hopkins, C. R.**, Epidermal growth factor and mitogenesis, *Nature (London)*, 286, 205, 1980.

104. **Marx, J. L.**, Tumour viruses and the kinase connection, *Science*, 211, 1336, 1981.

105. **Draghi, E., Armata, U., Andreis, G., and Mengato, L.**, The stimulation by epidermal growth factor (urogastrone) of the growth of neonatal rat hepatocytes in primary tissue culture and its modulation by serum and associated pancreatic hormones, *J. Cell Physiol.*, 103, 129, 1980.

106. **Lee, L.-S. and Weinstein, I. B.**, Tumor-promoting phorbol esters inhibit binding of epidermal growth factor to cellular receptors, *Science*, 202, 313, 1978.

107. **Shoyab, M., DeLarco, J. E., and Todaro, G. J.**, Biologically active phorbol esters specifically alter affinity of epidermal growth factor membrane receptors, *Nature (London)*, 279, 387, 1979.

108. **Driedger, P. E. and Blumberg, P. M.**, Specific binding of phorbol ester tumor promoters, *Proc. Natl. Acad. Sci. U.S.A.*, 77, 567, 1980.

109. **Roth, J., LeRoith, D., Shiloach, J., Rosenzweig, J. L., Lesniak, M. A., and Havrankova, J.,** The evolutionary origins of hormones, neurotransmitters, and other extracellular chemical messengers. Implications for mammalian biology, *N. Engl. J. Med.*, 306, 523, 1982.

110. **Paranjpe, M. S., DeLarco, J. E., and Todaro, G. J.,** Retinoids block ornithine decarboxylase induction in cells treated with tumour promoter TPA or the peptide growth hormones, EGF and SGF, *Biochem. Biophys. Res. Commun.*, 94, 586, 1980.

111. **Todaro, G. J., DeLarco, J. E., and Sporn, M. B.,** Retinoids block phenotypic cell transformation produced by sarcoma growth factor, *Nature (London)*, 276, 272, 1978.

112. **Rose, S. P., Stahn, R., Passovoy, D. S., and Hershman, H.,** Epidermal growth factor enhancement of skin tumor induction in mice, *Experientia*, 32, 913, 1976.

113. **Jetten, A. M.,** Retinoids specifically enhance the number of epidermal growth factor receptors, *Nature (London)*, 284, 626, 1980.

114. **Jetten, A. M.,** Action of retinoids and phorbol esters on cell growth and the binding of epidermal growth factor, *Ann. N.Y. Acad. Sci.*, 359, 200, 1981.

115. **Dicker, P. and Rozengurt, E.,** Retinoids enhance mitogenesis by tumour promoter and polypeptide growth factors, *Biochem. Biophys. Res. Commun.*, 91, 1203, 1979.

116. **Furstenberger, G., Berry, D. L., Sorg, B., and Marks, F.,** Skin tumor promotion by phorbol esters is a two-stage process, *Proc. Natl. Acad. Sci. U.S.A.*, 68, 7722, 1981.

117. **Cohlan, S. Q.,** Excessive intake of vitamin A as a cause of congenital anomalies in the rat, *Science*, 117, 535, 1953.

118. **Hardy, M. H., Dhouailly, D., and Sengel, P.,** Scales into feathers - an effect of retinoic acid on tissue interactions in the developing chick, *Ann. N.Y. Acad. Sci.*, 359, 394, 1981.

119. **Maden, K.,** Vitamin A and pattern formation in the regenerating limb, *Nature (London)*, 295, 672, 1982.

120. **Strickland, S. and Sawey, M. J.,** Studies on the effect of retinoids on the differentiation of teratocarcinoma stem cells *in vitro* and *in vivo*, *Dev. Biol.*, 78, 76, 1980.

121. **Strickland, S., Smith, K. K., and Marotti, K. R.,** Hormonal induction of differentiation in teratocarcinoma stem cells: generation of parietal endoderm by retinoic acid and dibutyryl CAMP, *Cell*, 21, 347, 1980.

122. **Plet, A., Evain, D., and Anderson, W. B.,** Effect of retinoic acid treatment of F9 embryonal carcinoma cells on the activity and distribution of cyclic AMP-dependent protein kinase, *J. Biol. Chem.*, 257, 889, 1982.

123. **Hughes, R. C. and Butters, T. D.,** Glycosylation patterns in cells: an evolutionary marker? *TIBS*, 6, 228, 1981.

124. **Morre, D. J., Creek, J. E., Morre, D. M., and Richardson, C. L.,** Glycosylation reactions and tumor establishment: modulation by vitamin A, *Ann. N.Y. Acad. Sci.*, 359, 367, 1981.

125. **Tickle, C., Alberts, B., Wolpert, L., and Lee, J.,** Local application of retinoic acid to the limb bud mimics the action of the polarizing region, *Nature (London)*, 296, 564, 1982.

126. **DeLuca, L., Kleinmann, H. K., Little, E. P., and Wolf, G.,** RNA metabolism in rat intestinal mucosa of normal and vitamin A-deficient rats, *Arch. Biochem. Biophys.*, 145, 332, 1971.

127. **Giesow, M.,** A carbohydrate signal for intracellular transit, *Nature (London)*, 290, 15, 1981.

128. **Snider, M. D.,** Transport of sugars during glycoprotein synthesis, *Nature (London)*, 298, 117, 1982.

129. **DeLuca, L. M., Silverman-Jones, C. S., Barr, R. M.,** Biosynthetic studies on mannolipids and mannoproteins of normal and vitamin A-depleted hamster livers, *Biochim. Biophys. Acta*, 409, 342, 1975.

130. **DeLuca, L. M., Bhat, P. V., Sasak, W., and Adamo, S.,** Biosynthesis of phosphoryl and glycosyl phosphoryl derivatives of vitamin A in biological membranes, *Fed. Proc. Fed. Am. Soc. Exp. Biol.*, 38, 2535, 1979.

131. **Dion, L. D., DeLuca, L. M., and Colburn, N. H.,** Phorbol ester-induced anchorage independence and its antagonism by retinoic acid correlates with altered expression of specific glycoproteins, *Carcinogenesis*, 2, 951, 1981.

132. **Frot-Coutaz, J., Letoublon, R., and Got, R.,** *In vitro* vitamin A-mediated glycosylation: recent developments in enzymatic studies, *Ann. N.Y. Acad. Sci.*, 359, 298, 1981.

133. **Quill, H. and Wolf, G.,** Formation of $\alpha,1,2$- and α-1,3-linked mannose disaccharides from mannosyl retinyl phosphate by rat liver membrane enzymes, *Ann. N.Y. Acad. Sci.*, 359, 331, 1981.

134. **Shidiji, Y., Sasak, W., Silverman-Jones, C. S., and DeLuca, L. M.,** Recent studies on the involvement of retinyl phosphate as a carrier of mannose in biological membranes, *Ann. N.Y. Acad. Sci.*, 359, 345, 1981.

135. **Wolf, G., Kiorpes, T. C., Masushige, S., Schreiber, J. B., Smith, M. J., and Anderson, R. S.,** Recent evidence for the participation of vitamin A in glycoprotein synthesis, *Fed. Proc. Fed. Am. Soc. Exp. Biol.*, 38, 2540, 1979.

136. **Wieland, T.,** Are sulfated cell surface glycoproteins involved in the control of cell differentiation? *TIBS*, 7, 308, 1982.

137. **Kiorpes, T. C. and Anderson, R. S.,** Underglycosylation of rat serum α_1-macroglobulin (α_1-M) in vitamin A deficiency, *Ann. N.Y. Acad. Sci.,* 359, 401, 1981.

138. **Olson, J. A., Rajanapo, W., and Lamb, A. J.,** Effect of vitamin A status on the differentiation and function of goblet cells in the rat intestine, *Ann. N.Y. Acad. Sci.,* 359, 181, 1981.

139. **Hassell, J. R. and Newsome, D. A.,** Vitamin A-induced alterations in corneal and conjunctival epithelial glycoprotein synthesis, *Ann. N.Y. Acad. Sci.,* 359, 358, 1981.

140. **Rosso, G. C., Bendrick, C. J., and Wolf, G.,** *In vivo* synthesis of lipid-linked oligosaccharides in the livers of normal and vitamin A-deficient rats, *J. Biol. Chem.,* 256, 8341, 1981.

141. **Rosso, G. C. and Wolf, G.,** Synthesis of lipid-linked oligosaccharides in vitamin A-deficient rat liver *in vivo, Ann. N.Y. Acad. Sci.,* 359, 418, 1981.

142. **Varandani, P. T., Wolf, G., and Johnson, B. C.,** Function of vitamin A in the synthesis of 3'-phosphoadenosine-5'-phosphosulfate, *Biochem. Biophys. Res. Commun.,* 3, 97, 1960.

143. **Shapiro, S. S. and Mott, D. J.,** Modulation of glycosaminoglycan biosynthesis by retinoids, *Ann. N.Y. Acad. Sci.,* 359, 306, 1981.

144. **Kojima, J., Nakomura, N., Kanatani, M., and Okiyama, M.,** Glycosaminoglycans in 3'-methyl-4-dimethylaminoazobenzene-induced rat hepatic cancer, *Cancer Res.,* 42, 2857, 1982.

145. **Taussig, M. J.,** *Processes in Pathology,* Blackwell Scientific, Oxford, 1979, 59.

146. **Pitot, H. C.,** *Fundamentals of Oncology,* 2nd ed., Marcel Dekker, New York, 1981, 220.

147. **Tomita, Y., Himene, K., Namato, K., and Endo, H.,** Combined treatments with vitamin A and 5-fluorouracil and the growth of allotransplantable and syngeneic tumors in mice, *J. Natl. Cancer Inst.,* 68, 823, 1982.

148. **Baron, S., Kleyn, K. M., Russell, J. K., and Blalock, J. E.,** Retinoic acid: enhancement of a tumor and inhibition of interferons antitumor action, *J. Natl. Cancer Inst.,* 67, 95, 1981.

149. **Felix, E. L., Lloyd, B., and Cohen, M. H.,** Inhibition of the growth and development of a transplantable murine melanoma by vitamin A, *Science,* 189, 886, 1975.

150. **Farber, E.,** Chemical carcinogenesis, *N. Engl. J. Med.,* 305, 1379, 1981.

151. **Slaga, T. J., Fischer, S. M., Weeks, C. E., Klein-Szanto, A. J. P., and Reiners, J.,** Studies on the mechanisms involved in multistage carcinogenesis in mouse skin, *J. Cell Biochem.,* 18, 99, 1982.

152. **Fusenig, N. E. and Samset, W.,** Growth-promoting activity of phorbolester TPA on cultured mouse skin keratinocytes, fibroblasts and carcinoma cells, in *Carcinogenesis: A Comprehensive Survey, Vol. 2, Mechanisms of Tumor Promotion and Cocarcinogenesis,* Slaga, T. J., Sivak, A., and Boutwell, R. K., Eds., Raven Press, New York, 1978, 203.

153. **Miller, J. A.,** Carcinogenesis by chemicals: an overview. GHA Clowes Memorial Lecture, *Cancer Res.,* 30, 559, 1970.

154. **Miller, E. C.,** Some current perspectives on chemical carcinogenesis in humans and experimental animals: Presidential Address, *Cancer Res.,* 38, 1479, 1978.

155. **Ya Kon, I. and Shirina, L. I.,** Vitamin A and NADP-dependent isocitric dehydrogenase: differences in action on mitochondrial and cytoplasmic isoenzyme fractions, *Biochemistry (USSR),* 46, 647, 1981.

156. **Colby, H. D., Kramer, R. E., Greiner, J. W., Robinson, D. A., Krause, R. F., and Canady, W. J.,** Hepatic drug metabolism in retinol-deficient rats, *Biochem. Pharmacol.,* 24, 1644, 1975.

157. **Hauswirth, J. W. and Brizvela, B. S.,** The differential effects of chemical carcinogens on vitamin A status and on microsomal drug metabolism in normal and vitamin A-deficient rats, *Cancer Res.,* 36, 1941, 1976.

158. **Kalamegham, R. and Krishnaswamy, K.,** Benzo(a)pyrene metabolism in vitamin A-deficient rats, *Life Sci.,* 27, 33, 1980.

159. **Siddik, Z. H., Drew, R., Litterst, C. L., Minnaugh, E. G., Sikic, B. I., and Gram, T. E.,** Hepatic cytochrome P-450-dependent metabolism and enzymatic conjugation of foreign compounds in vitamin A-deficient rats, *Pharmacology,* 21, 383, 1980.

160. **Hill, D. L. and Shih, T.-W.,** Vitamin A compounds and analogs as inhibitors of mixed-function oxidases that metabolize carcinogenic polycyclic hydrocarbons and other compounds, *Cancer Res.,* 34, 564, 1974.

161. **Dickens, M. S. and Sorof, S.,** Retinoid prevents transformation of cultured mammary glands by procarcinogens but not by many activated carcinogens, *Nature (London),* 285, 581, 1980.

162. **Sirianni, S. R., Chen, H. H., and Huang, C. C.,** Effect of retinoids on plating efficiency, sister chromatid exchange (SCE) and mitomycin-C-induced SCE in cultured Chinese hamster cells, *Mutat. Res.,* 90, 175, 1981.

163. **Busk, L. and Ahlborg, U. G.,** Retinol (vitamin A) as a modifier of 2-amino fluorene and 2-acetylaminofluorene mutagenesis in the salmonella/microsome assay, *Arch. Toxicol.,* 49, 169, 1982.

164. **Baird, M. B. and Birnbaum, L. S.,** Inhibition of 2-fluorenamine-induced mutagenesis in *Salmonella typhimurium* by vitamin A, *J. Natl. Cancer Inst.,* 63, 1093, 1979.

165. **Malathi, S. and Ganguly, J.,** Studies on the metabolism of vitamin A_7. Lowered biosynthesis of ascorbic acid in vitamin A-deficient rats, *Biochem. J.,* 92, 521, 1964.

166. **Treschel, U., Fleisch, H., and Manod, A.**, Retinol and retinoic acid modulate the metabolism of 25-hydroxy vitamin D_3 in kidney cell culture, *FEBS Lett.*, 135, 115, 1981.

167. **Alcantara, E. N. and Speckmann, E. W.**, Diet, nutrition, and cancer, *Am. J. Clin. Nutr.*, 29, 1035, 1976.

168. **Mathews-Roth, M. M.**, Antitumour activity of β-carotene, canthaxanthin and phytoene, *Oncology*, 39, 33, 1982.

169. **McCormick, D. L., Mehta, R. G., Thompson, C. A., Dinger, N., Caldwell, J. A., and Moon, R. C.**, Enhanced inhibition of mammary carcinogenesis by combined treatment with N-(4-hydroxyphenyl)retinamide and ovariectomy, *Cancer Res.*, 42, 508, 1982.

170. **Nettesheim, P. and Williams, M. L.**, The influence of vitamin A on the susceptibility of the rat lung to 3-methylcholanthrene, *Int. J. Cancer*, 17, 351, 1976.

171. **Schmähl, D., Kruger, C., and Preissler, P.**, Versuche zur krebsprophylaxe mit vitamin A, *Arzneimittelforschung*, 22, 946, 1972.

172. **Wenk, M. L., Ward, J. M., Reznik, G., and Dean, J.**, Effects of three retinoids on colon adenocarcinomas, sarcomas and hyperplastic polyps induced by intrarectal N-methyl-N-nitrosourea administration in male F344 rats, *Carcinogenesis*, 2, 1161, 1981.

173. **Daoud, A. H. and Griffin, A. C.**, Effect of retinoic acid, butylated hydroxytoluene, selenium and sorbic acid on azo-dye hepatocarcinogenesis, *Cancer Lett.*, 8, 299, 1980.

174. **Verma, A. K., Conrad, E. A., and Boutwell, R. K.**, Inhibition of mouse skin carcinogenesis by a retinoid, steroid, and a protease inhibitor, *Proc. Am. Assoc. Cancer Res.*, 21, 93, 1980.

175. **Weeks, C. E., Slaga, T. J., Hennings, H., Gleason, C. L., and Bracken, W. M.**, Inhibition of phorbol ester-induced tumor promotion in mice by vitamin A analog and anti-inflammatory steroid, *J. Natl. Cancer Inst.*, 63, 401, 1979.

176. **Ip, C. and Ip, M. M.**, Chemoprevention of mammary tumorigenesis by a combined regimen of selenium and vitamin A, *Carcinogenesis*, 2, 915, 1981.

177. **Thompson, H. J., Meeker, L. D., Tagliaterro, A. R., and Becci, P. J.**, Effect of retinylacetate on the occurrence of ovarian hormone-responsive and non-responsive mammary cancers in the rat, *Cancer Res.*, 42, 903, 1982.

178. **McCormick, D. L. and Moon, R. C.**, Influence of delayed administration of retinyl acetate on mammary carcinogenesis, *Cancer Res.*, 42, 2639, 1982.

179. **Thompson, H. J., Meeker, L. D., Tagliaferro, A. R., and Becci, P. J.**, Effect of retinyl acetate on the occurrence of ovarian hormone-responsive and non-responsive mammary cancers in the rat, *Cancer Res.*, 42, 903, 1982.

180. **McCormick, D. L., Mehta, R. G., Thompson, C. A., Dinger, N., Caldwell, J. A., and Moon, R. C.**, Enhanced inhibition of mammary carcinogenesis by combined treatment with N-(4-hydroxyphenyl) retinamide and ovariectomy, *Cancer Res.*, 42, 508, 1982.

181. **Cohen, L. A. and van der Snoek, L.**, Concentration-dependent stimulation/inhibition of mammary tumor cell growth by retinyl acetate *in vitro*, *Proc. Am. Assoc. Cancer Res.*, 22, 456, 1981.

182. **Maiorana, A. and Gullino, P. M.**, Effect of retinyl acetate on the incidence of mammary carcinomas and hepatomas in mice, *J. Natl. Cancer Inst.*, 64, 655, 1980.

183. **Mettlin, C. and Graham, S.**, Dietary risk factors in human bladder cancer, *Am. J. Epidemiol.*, 110, 255, 1979.

184. **Gerschaff, S. N. and McGandy, R. B.**, Effects of vitamin A-deficient diets containing lactose in producing bladder calculi and tumors in rats, *Am. J. Clin. Nutr.*, 34, 483, 1981.

185. **Mettlin, C., Graham, S., and Swanson, M.**, Vitamin A and lung cancer, *J. Natl. Cancer Inst.*, 62, 1435, 1979.

186. **Mossman, B. T., Craighead, J. E., and Macpherson, B. V.**, Asbestos-induced epithelial changes in organ cultures of hamster trachea: inhibition by retinyl methyl ether, *Science*, 207, 311, 1980.

187. **Sporn, M. B. and Newton, D. L.**, Retinoids and chemoprevention of cancer, in *Inhibition of Tumor Induction and Development*, Zedeck, M. S. and Lipkin, M., Eds., Plenum, New York, 1981, 71.

188. **Wertz, P. W., Kensler, T. W., Mueller, G. C., Verma, A. K., and Boutwell, R. K.**, 5,6-Epoxyretinoic acid opposes the effects of 12-0-tetradecanoylphorbol-13-acetate in bovine lymphocytes, *Nature (London)*, 277, 227, 1979.

189. **Dion, L. D. and Gifford, G. E.**, Vitamin A-induced modulation of the transformed cell phenotype *in vitro*, *Ann. N.Y. Acad. Sci.*, 359, 389, 1981.

190. **Yamamoto, N., Bister, K., and Zurhausen, H.**, Retinoic acid inhibition of Epstein-Barr virus induction, *Nature (London)*, 278, 553, 1979.

191. **Atukuroka, S., Basu, T. K., Dickerson, J. W. T., Donaldson, D., and Sakula, A.**, Vitamin A, zinc and lung cancer, *Br. J. Cancer*, 40, 927, 1979.

192. **Wald, N., Idle, M., Boreham, J., and Bailey, A.**, Low serum vitamin A and subsequent risk of cancer. Preliminary results of a prospective study, *Lancet*, 2, 813, 1980.

193. Editorial, Oral contraceptives and breast neoplasia, *Br. Med. J.*, 1, 545, 1976.
194. **Basu, T. K., Donaldson, D., Jenner, M., Williams, D. C., and Sakula, A.**, Plasma vitamin A in patients with bronchial carcinoma, *Br. J. Cancer*, 33, 119, 1976.
195. **Slaga, T. J., Fischer, S. M., Viaje, A., Berry, D. L., Bracken, W. M., LeClerc, S., and Miller, D. R.**, Inhibition of tumor promotion by antiinflammatory agents: an approach to the biochemical mechanism of promotion, in *Carcinogenesis: A Comprehensive survey, Vol. 2. Mechanisms of Tumor Promotion and Cocarcinogenesis*, Slaga, T. J., Sivak, A., and Boutwell, R. K., Eds., Raven Press, New York, 1978, 173.
196. **Reddy, T. V. and Weisburger, E. K.**, Hepatic vitamin A status of rats during feeding of the hepatocarcinogen 2-aminoanthroquinone, *Cancer Lett.*, 10, 39, 1980.
197. **Borek, C., Miller, R. C., Geard, C. R., Guernsey, D., Osmak, R. S., Rutledge-Freeman, M., Ong, A., and Mason, M.**, The modulating effect of retinoids and a tumor promoter on malignant transformation sister chromatid exchanges, and Na/K ATPase, *Ann. N.Y. Acad. Sci.*, 359, 237, 1981.
198. **Birt, D. F., Sayed, S., Davies, M. H., and Pour, P.**, Sex differences in the effects of retinoids on carcinogenesis by N-nitrosobis (2-oxopropyl) amine in Syrian hamsters, *Cancer Lett.*, 14, 13, 1981.
199. **Davies, R. E., Forbes, P. D., and Urbach, F.**, Promotion of carcinogenesis by all-*trans*- retinoic acid (RA): effect of initiator, *Proc. Am. Assoc. Cancer Res.*, 22, 374, 1981.
200. **Welsch, C. W., Goodrich-Smith, M., Brown, C. K., and Crowe, N.**, Enhancement by retinyl acetate of hormone-induced mammary tumourigenesis in female GR/A mice, *J. Natl. Cancer Inst.*, 67, 935, 1981.

Chapter 6

APPLICATION OF THE ERYTHROCYTE GLUTATHIONE REDUCTASE TEST TO THE EVALUATION OF VITAMIN B₂ STATUS IN NORMAL FEMALE SUBJECTS AND IN FEMALES TAKING ORAL CONTRACEPTIVE AGENTS

Suzanne Y. Tonkin

TABLE OF CONTENTS

I. REVIEW OF THE LITERATURE

A. General Introduction

The combined oral contraceptive pill has proven to be a reliable means of preventing conception, but it becomes a less attractive method when there are concomitant adverse side effects. The latter is expressed in many different ways and one of these, a change in vitamin status, was investigated.

The component synthetic steroids are estrogen and progestogen analogues whose properties ensure that an effective concentration reaches receptors of target tissues, and thus mediate contraceptive actions. However, their own metabolism produces numerous other compounds which, together with the parent compound, are responsible for secondary metabolic changes. Since ovulation control via this method can only be achieved in that milieu, it is desirable that the biochemical changes be documented and that any clinical sequelae should be evaluated. At best these can be minimized by varying the steroid content with respect to dose level and chemical type.

It is apparent from cumulative evidence of both body chemistry parameters and clinical signs, that these orally active steroids do elicit side effects. Prominent among the many adverse consequences which have been documented are cardiovascular changes. The list of recognized risk factors for arterial thrombosis, especially myocardial infarction, and venous thrombosis, includes oral contraception and these factors act synergistically. Women who would be particularly likely to incur unfavorable side effects can be identified from their medical history (for example, present or past disease of the circulatory system, liver diseases, carcinoma of the breast), and for them the Pill would be absolutely contraindicated. For those with strong relative contraindications such as diabetes, conditions which would be aggravated by fluid retention, and severe depression, the Pill should only be used under special supervision. Serious side effects would therefore be anticipated in these subjects. However, a proportion of women who are judged to be outside of these risk classifications will also experience side effects, and their unpredictable occurrence is probably due to the considerable individual variations in pharmacological response to the hormones. It would appear that a subgroup of healthy women are more susceptible to this type of medication and react accordingly. If such problems do not recede after a few cycles, an alternative is to change to another prescription, and schemes for selecting estrogen and/or progestogen type and dose have been devised for medical practitioners to follow based on the actual symptoms.

Clinical experience was gained primarily with higher doses of steroids than are currently used. Estrogens had been implicated in some of the more serious problems that could arise and, since these appeared to be dose related, it was believed necessary to produce a Pill which had the minimum amount of estrogen compatible with contraceptive efficacy. This was achieved with 50 and now 30 μg ethynylestradiol. Since the inception of the Pill, both the progestogen and estrogen contents have been reduced severalfold. Viewed in an historical perspective, 50 μg was a low dose, and the difference between 30 and 50 μg might seem relatively insignificant. However, in many respects 50 μg pills did not alleviate problems and a lower dose (30 μg) was tried, but a more powerful progestogen component is required for the Pill to remain effective. One progestogen believed to be particularly promising was levo-norgestrel (d-NG) since it was claimed to be more potent as a contraceptive compound and could therefore be taken at a lower dose level than other progestogens such as norethisterone. The combination of 30 μg ethynylestradiol with 150 μg d-NG therefore provided a low-dose pill with respect to both of the steroid components and it was expected to have favorable metabolic and clinical implications. Although the estrogen component was the primary consideration, the lowest dose was only marketed with d-NG.

At the time this study was contemplated, a Pill containing the lowest dose of estrogen received substantial patronage, but its impact on vitamin status and most other aspects of health was not known. Therefore, a comparative study was instigated to evaluate combined oral contraceptives with the lowest estrogen content, i.e., 30 μg ethynylestradiol, 150 μg d-NG, hereafter referred to as a low-dose Pill.* In the study presented, this type was distinguished from other prescriptions which contained at least 50 μg estrogen and 250 μg progestogen (high-dose Pill). A majority of the volunteers were recipients of these two Pill categories, and could be classified thus, but this distinction between lower and higher dose pills became less clear with the introduction of other combinations as well as ones in which the regime changed during a cycle.

The metabolic basis of various clinical manifestations has been researched extensively, but is open to speculation as there are likely to be many contributing factors. Some aspects of biochemistry and metabolism which have been studied encompass proteins and amino acids, enzymes, trace elements, vitamins, hormones, carbohydrates, and lipids and it has been shown that oral contraceptives can induce numerous changes in body chemistry. It is therefore difficult to link particular biochemical and clinical abnormalities. Since vitamins have fundamental roles in biochemical processes, there is justification for discovering whether and how oral contraceptives affect them. A vitamin deficiency can be caused by an inadequate dietary intake but may also develop despite a satisfactory diet under circumstances which reduce the effective concentration or increase the demands of the body. For example, certain exogenous medication and medical conditions can interfere with vitamin status in these ways and it is important to know whether oral contraceptives have this influence.

The nutritional status of women who relied on orally administered steroid contraceptive compounds has been the subject of many investigations involving higher dose pills (i.e., ≥50 μg estrogen) and a review of this literature has already been undertaken.[1] All of the vitamins for which information was available are relevant, because of the interrelationships and cooperation between nutrients in metabolism, but the research enquiry was restricted. For some vitamins there was agreement that a measurable change took place, but the

---

* The term "low dose" describes oral contraceptives containing 30 μg ethynylestradiol per 150 μg d-NG. The term "high dose" describes oral contraceptives containing >30 μg ethynylestradiol per >250 μg of a progestogen.

metabolic basis was obscure and open to speculation. Inconsistent reports for the effect on other vitamins await elucidation.

A change in vitamin status could be clinically significant, whether the women are already malnourished or have good nutritional status. Previous studies were conducted in women who represented this spectrum. An investigation that links vitamins and oral contraceptives is of particular relevance to the inhabitants of regions wherein poor diet, disease, and treatment with other drugs commonly coexist. Each of these has the potential to reduce vitamin status, and oral contraceptives should not be aggravating. Changes in vitamin status might also have clinical significance in well-nourished women who were not exposed to these risks, since variations could be linked to a number of adverse side effects, including for example, fetal malformations (vitamin A), psychiatric symptoms, impaired glucose tolerance, cancer of the urinary tract (vitamin B_6), megaloblastic anemia (vitamin B_{12} and folate), and cardiovascular disease (vitamins C and K). There might be less disturbance from a lower dose combination of steroids.

While it would be desirable to measure most of the vitamins, limitations had to be imposed and therefore three B vitamins were studied (B_1, B_2, and B_6). Vitamin B_2 was interesting because it was apparent from the review that higher dose oral contraceptives had an adverse influence. This is especially undesirable in the regions where a deficiency is already endemic. If the lower dose Pill offered an advantage, then it should be demonstrable.

The evaluation of vitamin status in people who inhabit a region where avitaminosis is rarely encountered is best made by using a biochemical approach. This also applies to the assessment of drug-induced changes in nutrient levels, since experience has shown that the type of change usually encountered is difficult to define by any other means. Functional enzymatic tests, applicable to thiamin, riboflavin, and pyridoxine by virtue of a metabolic role executed by coenzymes, were particularly suitable for the type of investigation contemplated. Furthermore, such methods had been developed as part of an intention to encourage the application of a limited number of techniques with the attendant advantages of uniformity and standardization, a task which was originated and coordinated by a group within the World Health Organization (WHO).

It was necessary to define certain aspects of the test used to quantitate vitamin B_2 in the reference sample, namely the erythrocyte glutathione reductase test. Before applying these in the field, preliminary work was directed at laboratory aspects. The results were considered in three sections, covering laboratory aspects of the test, its application to a reference sample, and also to the evaluation of vitamin status in females taking oral contraceptives. The impact of oral contraceptive agents on vitamin status was studied by examining two samples which were drawn from the same population, using the cross-sectional approach.

Studies which are carried out over an extended period require an awareness of variables related to the passage of time. Dietary vitamin sources fluctuate according to seasonal variation, and this may need to be taken into consideration. However, the opinion of experts questioned was that the three vitamins under investigation would be uniformly available throughout the year in the region under study.

The results obtained from the reference sample were examined in some detail. One reason this was deemed necessary was that many of the studies on which the application of biochemical tests has been based were carried out under experimentally controlled circumstances. While these were valuable, they may not represent a stable nutritional state which could be encountered naturally. It was therefore judged relevant to determine whether the generally accepted conclusions from depletion-repletion studies could be applied to a nutritionally stable sample. Therefore, instead of measuring an individual on many occasions, giving rise to a range of presymptomatic states, a group of people whose levels covered a range was measured once. Among several matters considered was the relationship between basal enzymatic activity and the index activity coefficient, in view of the reliance which is placed on the latter for interpreting these types of tests. To base conclusions on the ratio required an assurance that it was a valid measurement of vitamin status in this type of study. Therefore the outcome of these analyses had implications for the comparison of a test and reference sample using the most suitable variable(s).

Several vitamin interrelationships have been recognized, therefore if the effect of a drug is to alter the level of one or more nutrients, the sequelae may also be expressed in other alterations which are actually secondary. While it was possible to discover whether ingestion of oral contraceptives was reflected in a biochemical change to each vitamin considered separately, it may be difficult to distinguish between those changes in vitamin status which are the direct result of oral contraceptive use, and those which occur indirectly via an intermediate change. This aspect was therefore explored separately (unpublished observations) by correlating the results of each of the vitamin assays for each female volunteer.

B. Metabolism and Metabolic Function

The formation of flavin coenzymes mediated by ATP is shown in Figure 1. Free riboflavin, mainly from ingested food, but partly from synthesis by intestinal flora is understood to be absorbed in the small intestine, converted to flavin mononucleotide (FMN) in the intestinal mucosa, and then to flavin adenine dinucleotide (FAD) by the liver.[2] Such patterns of absorption, transport, and excretion have been recently described,[3] and the fate of riboflavin in the mammal was the subject of another review.[4] Most of the riboflavin in the tissues was found stored in the form of FAD and FMN, although a little remained free. In rat liver, this was estimated to be about 2% of the total, and FMN represented about 16%.[3] The transport, distribution, and elimination of a vitamin can be determined by the extent of its binding to plasma proteins. Once it was absorbed from the intestinal tract, riboflavin and FMN were located bound to alpha and beta globulins[5] and to albumin.[6] However, the plasma protein binding was not very extensive, making its effect on distribution and elimination small. The affinity of FMN was much greater for certain proteins in the tissues,[6] wherein the flavin coenzymes combined with these proteins to form catalytically active flavoproteins.

B_2 appears to be unique in that it is not extensively decomposed by the body, and therefore degradative products, apart from those formed by bacteria, were not detected in the urine.[7] Free vitamin was excreted and thus urinary excretion was shown to correlate well with reserves and intake. When a large dose was administered, nearly all of it was excreted. The maintenance requirement appeared to be largely dependent upon excretion rather than decomposition.

The metabolic role of riboflavin mainly resides in two cofactors, FMN and FAD. The ingested flavin is retained as a cofactor in association with enzymes in order to function catalytically in biological oxidation-reduction reactions. It was judged inappropriate to attempt even a synopsis of the flavoproteins since there are so many and their functions are quite diverse.[8] Comprehensive discussions of this class of enzyme can be found in most biochemistry texts. There are many more enzymes known to require FAD than FMN.

Most of the flavoproteins are located in the mitochondria and would not exist in erythrocytes. One of the exceptions merits further attention since it is the basis for a biochemical test for the assessment of vitamin B_2 status (see Section I.E.). GR or glutathione reductase (E.C.1.6.4.2) is one of two flavoproteins reported to be found in the erythrocytes and which require FAD,[9] the other flavoprotein being reduced nicotinamide adenine dinucleotide (NADH)-methemoglobin reductase. Recently an NADPH-methemoglobin flavin reductase was described.[10] Most of the riboflavin in blood is believed to be in the form of FAD, and GR is the major flavoprotein in red cells. Its activity is taken to be a reflection of the amount of riboflavin in the tissues since it shows the intracellular availability of the coenzyme to an enzyme which has a fundamental role. The enzyme belongs to a sub-class of flavoprotein

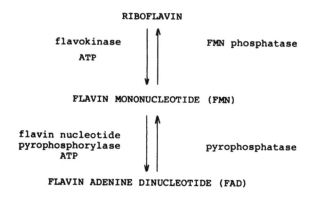

FIGURE 1.   Formation of flavin coenzymes.

which promotes the reduction of a disulfide bond to two sulfydryl groups utilizing NADPH. GR catalyzes the step:

$$\underset{\text{oxidized glutathione}}{\text{G-S-S-G}} + \underset{\text{(NADH)}}{\text{NADPH} + \text{H}^+} \xrightarrow[\text{reductase}]{\text{glutathione}} \underset{\text{glutathione}}{\text{2 G-SH}} + \underset{\text{(NAD}^+\text{)}}{\text{NADP}^+}$$

Glutathione is a tripeptide, γ-glutamyl-cysteinyl-glycine. It has been shown that GR is a flavin enzyme with FAD serving as a prosthetic group.[11] However, there was a reactive disulfide group present, apparently adjacent to the flavin, which also participated in the chemical reaction.[12]

C. Effect of Riboflavin Deficiency on Metabolism

Because the flavin coenzymes FMN and FAD are involved in many facets of intermediary metabolism, a deficiency is manifest in numerous processes. Some of these were considered in a review of riboflavin metabolism in health and disease.[3] The sequence of events which is believed to take place during depletion begins with circulating riboflavin. Protein-bound forms are then affected and the tissue vitamin, as FMN or FAD, is mobilized. However, in a deficient state, the body seems to preserve a minimum level of riboflavin-containing enzymes by compensatory weight loss, and newer sources are needed for tissue replacement or repair.[13] The retention of physiological amounts of riboflavin in the body was judged to be mainly due to its high affinity for apoflavoproteins.[6]

Flavin-requiring enzymes were markedly affected by a decrease in availability of their coenzyme, according to hepatic analyses,[3] although different responses were noticed between enzymes. The decreased concentrations of hepatic riboflavin, FAD, FMN, and the activities of FMN- and FAD-requiring enzymes were readily demonstrated in animals and FMN decreased proportionately more than FAD.[13] Although it was not surprising that these changes occurred under limiting conditions, since the vitamin is the direct precursor of the two coenzymes, it has been suggested that the pathogenesis of the deficiency state may be more complex.[3] The demonstration of lowered activity of hepatic flavokinase and concomitantly less enzymatic synthesis of FMN would also account for the lowered flavin levels. Furthermore, increased activity of FAD pyrophosphorylase would promote FAD synthesis at the expense of FMN (Figure 1). These changes had the effect of conserving FAD at the expense of FMN, and since the demand for FAD as a cofactor to enzymes was greater than for FMN, this represented utilization of riboflavin for its most essential functions when the supply of the vitamin was limited.

There are other complex sequelae involving the metabolism of another vitamin. The biochemical interrelationship between riboflavin and pyridoxine is believed to be the underlying explanation for observations pertaining to the biochemical tests and the clinical condition in vitamin B_2 and B_6 deficiencies[14] because pyridoxine phosphate oxidase is FMN dependent. Pyridoxine (pyridoxamine) phosphate oxidase catalyzes the synthesis of pyridoxal phosphate from pyridoxine phosphate or pyridoxamine phosphate.

D. Clinical Aspects

The metabolic effects of riboflavin deficiency are complex and widely expressed, thus it was expected that few of these events could be easily detected clinically. Although certain signs and symptoms have been characteristically observed in a state of deficiency,[15,16] the condition was difficult to describe from naturally occurring cases because it was rarely found in isolation, but rather in conjunction with other vitamin deficiencies. In addition, similar signs and symptoms exist during shortages of vitamins other than B_2 and the cooperation of several nutrients may be responsible, at least in part, for the molecular basis of a lesion. It was studies in which experimental vitamin B_2 deficiency was created in human volunteers that led to the recognition of clinical sequelae. In outline, the effects of restriction are manifest externally in the form of skin lesions such as inflammation of the mouth and a characteristic splitting lesion at the corners. Later, some changes were seen in bone marrow, apparent in a normochromic, normocytic anemia. A detailed description of the clinical aspects of experimental human riboflavin deficiency can be found in a recent text devoted to riboflavin.[17] Gross and cellular pathology of riboflavin-deficient animals have been documented.[3] The relatively minor nature of the lesions is at variance with the fundamental processes which rely on B_2 and this may be rationalized in terms of conservation and maintenance of the flavin content at a stable lower limit (see Section I.C.). It may only become severe during protein deficiency because these cofactors are generally tightly bound to the enzymes.

Clinical recognition of vitamin B_2 deficiency was particularly difficult in practice where an uncomplicated, single nutrient deficiency is the exception rather than the rule. A deficiency has been implicated in oral lesions such as angular stomatitis, glossitis, and cheilosis, but it may not be the cause.[14] In order to make a diagnosis and implement correct treatment, these studies have been aided by biochemical investigations. It was found that patients who exhibited clinical evidence of riboflavin deficiency improved markedly with pyridoxine supplementation.[18] Biochemical tests showed that riboflavin deficiency existed, but according to their clinical response, it was apparent that this was in addition to pyridoxine deficiency. In a related study[19] a group of patients who had orolingual lesions differed in their clinical response to a course of treatment in which B_2 was followed by B_6. Some of these responded best to B_2 and required additional B_6 for complete clearance of residual signs, whereas others required mainly B_6 and exhibited minimal response to riboflavin. A biochemical test indicated that poor vitamin B_2 status existed initially, apparent from the low GR activity. It was thought that among sufferers of angular stomatitis and glossitis there are two groups of patients: one predominantly deficient in vitamin B_2, the other lacking in B_6, and therefore the clinical response to one of these vitamins could be satisfactory.

It was postulated that the biochemical basis of the oral lesions was a cellular deficiency of pyridoxal phosphate, either due to dietary deficiency of pyridoxine or conditioned by inefficient enzymatic conversion of dietary pyridoxine to pyridoxal phosphate as a result of riboflavin deficiency.[18] On administration of B_6, the stores of B_2 may be utilized for formation of pyridoxal phosphate, resulting in a clinical cure, or pyridoxal phosphate may be formed through a different pathway which was not riboflavin dependent.[19] Corresponding changes were found in biochemical indexes of riboflavin status, but whether the demand created by a single flavin enzyme would have these effects was considered a moot point.[18] However, subsequent work showed that pyridoxal phosphate synthesis was affected.[20]

E. Laboratory Assessment of Vitamin B_2 Status

In view of the widespread metabolic function of riboflavin as coenzyme and the nature of the signs and symptoms of vitamin B_2 deficiency, it is difficult to make a diagnosis on the basis of these factors alone, even in cases uncomplicated by other deficiencies. Sub-clinical changes are virtually asymptomatic, and an investigation such as the one in this study would probably not be contemplated. Dietary analyses sometimes augment the clinical data, but it could not be assumed that a theoretically adequate diet is actually sufficient, since other factors such as cooking affect the nutrients and each individual has different requirements. Furthermore, reliable dietary intake information is difficult to obtain and although such data can provide general knowledge concerning the provision of riboflavin, it fails to define the body reserves and metabolic state. The influence of drugs such as oral contraceptives could not be assessed in this manner, and herein lies the value of laboratory procedures for the assessment of nutritional status. Since biochemical changes normally precede clinical changes, they could be applied to advantage in sub-clinical cases and in situations where a nondietary difference in B_2 status is suspected.

Assays which measure the growth response of animals such as rats and chickens are occasionally used, but these biological procedures have been largely superseded by others.[17] One of these is based on microbial growth in response to a nutrient. Usually a lactic acid-producing organism such as *Lactobacillus casei,* dependent on a supply of riboflavin from the medium, was grown and the lactic acid was measured. The microbiological assay has been viewed as unreliable in some hands, but it has also been compared favorably[21] with a chemical technique of measuring these compounds based on fluorimetry. Others involving radioisotopes have been developed, and if necessary the individual flavins can be separated by chromatography.[22]

It was only relatively recently that sensitive methods were developed to measure a vitamin B_2-dependent enzyme in order to assess the nutritional state. Until then the most commonly employed biochemical procedure was to measure urinary riboflavin excretion, although some data has also accrued for the concentration in blood.[23,24] The latter was based largely on the work of Bessey and co-workers.[25]

Direct measurement of the vitamin in blood, or its components, was usually done by fluorimetric or microbiological techniques. The concentration of riboflavin in erythrocytes decreased during restricted intake of the vitamin,[25] with a lesser effect being noticed in plasma. However, according to the experience in several laboratories, it was concluded that an analysis of riboflavin in blood was of limited value.[23] The concentration of free vitamin in serum was influenced by very recent intake and was too variable to serve as a useful index, whereas a criticism of the red cell riboflavin level was that it did not change until after more easily determined parameters had altered. However, because of this slower response erythrocyte riboflavin was believed to be a good index of tissue stores.[24,26] Restoration to the normal level was also slow,[25] suggesting that the stores were slowly filled after deprivation and were less susceptible to dietary influences.[24]

It was recognized that a single nutrient deficiency rarely occurred and that the coexistence of more than one could make the interpretation of laboratory test results difficult because of the influences some nutrients have upon others. Therefore, behavior of each vitamin should not be considered independently. In this context, the relationship between vitamins B_2 and B_6 was reflected in a higher erythrocyte riboflavin concentration during B_6 deficiency.[18] These results prompted the suggestion that erythrocyte riboflavin concentration was a poor index of B_2 status when there was associated B_6 deficiency.

Guidelines for the interpretation of erythrocyte riboflavin concentration have been defined,[23] but these were considered invalid when applied to the results from investigators in India[18,26] wherein clinical signs and symptoms coexisted with apparently normal levels, and there was no correlation between the clinical and biochemical results. It may therefore be necessary for each investigator to establish his own guidelines. While practical for the research laboratory, the application of this test to nutrition surveys was considered a daunting prospect,[24] but at that time the only alternative was to measure the urinary excretion since the enzymatic test had not been devised.

Considerably more information exists concerning the urinary excretion of riboflavin,[23,24,27] which could be measured equally well microbiologically or fluorimetrically. In contrast to most of the other vitamins, riboflavin is fairly stable in tissue (Section I.B.) and therefore the relationship between intake and excretion was considerably simpler than, for example, thiamin, and the B_2 excreted in the urine correlated with reserves and intake. Based on a number of carefully controlled human studies, as well as population surveys, this criterion of B_2 status has been well-defined. Less than 100 μg excreted each day was indicative of a recent dietary regimen providing less than the minimum requirement, and 1.3 to 1.6 μg/day was required to maintain tissue saturation. Clinical signs of a deficiency were associated with less than 50 μg excreted per day.

Notwithstanding this information, one of the main problems is to obtain a representative specimen. Since the urinary B_2 level tends to reflect the recent dietary intake, it is prone to variation, which could be augmented by the sampling method. A 24-hr specimen was recommended, but this would be difficult to obtain under many circumstances, and consequently random samples were often taken. To relate the values to creatinine may not be a solution since this could also vary markedly.[24,27] Misleading results were more likely to occur in individuals who had an adequate intake since those who had depleted body stores would retain rather than excrete any riboflavin ingested.[23] Another disadvantage was the influence on urinary riboflavin of other factors such as physical activity, nitrogen balance, temperature, and stress.

Riboflavin decreased soon after the dietary level was restricted during controlled depletion/repletion studies,[28] but because a low level persisted throughout the remainder of the depletion phase and well into the repletion stage, this response limited evaluation of the severity of ariboflavinosis. Urinary riboflavin did not correlate with the extent of the vitamin deficiency in the absence of clinical parameters and this may have been overlooked in studies of short duration.[28] Urinary analysis of this B vitamin was judged useful for estimating dietary intakes in populations prior to enzymological developments; as an individual diagnostic aid it was usually less informative.[24,29] A variety of factors could induce fluctuations in the excretion rate of water-soluble substances. These became less important when large numbers of individuals were considered.[24]

Balance studies using radioactively labelled vitamins, whereby input was equated with output, could detect the catabolic products of vitamin metabolism and for some vitamins such as thiamin, the measurement of metabolites might permit a more accurate assessment of nutritional status.[24] However, the approach was not suitable for riboflavin since this was the principal urinary excretion product, together with relatively small quantities of other metabolites.

One approach to the employment of urinary excretion of riboflavin which obviated some of its limitations is the load retention test.[23] In principle, controlled administration of vitamin B_2 and retrieval should reveal a degree of retention in keeping with the state of the tissues. This technique has received only limited attention and therefore the conclusions are tentative. The recommendation[23] that it should be interpreted in conjunction with other information may have discouraged its use.

Whereas urinary or plasma vitamin B_2 concentration seemed to reflect the previous days diet and decreased after restriction, protein-bound B_2 was not markedly affected because in tissue, as FMN or FAD, it was not mobilized until the circulating riboflavin had been depleted. A preferred assay would therefore be direct estimation of the protein-bound vitamin.

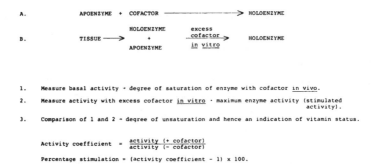

FIGURE 2.   Enzymatic functional test for the assessment of vitamin status.

The function of B_2 as a coenzyme component is well known, thereby allowing the development of an enzymatic functional test.

Two other vitamins for which a coenzymatic role has been ascribed, riboflavin and pyridoxine, can also be evaluated by enzymatic functional tests. The concept on which these are based can be stated succinctly in the manner of Figure 2 but this needs elaboration, since the practical interpretation of results from the test may be more complicated. The basal activity is regarded as a measure of the amount of enzyme which is catalytic in vivo. Since this depends on the coenzyme, the activity could also be proportional to the concentration of coenzyme in the cell. However, the activity is a measure of the intracellular availability of the coenzyme for one particular enzyme rather than the total concentration, and the relationship between these two amounts would need to be determined for each case. It has been recognized that pyridoxal phosphate in erythrocytes binds extensively to various proteins for which it has no coenzymatic role, whereas for thiamin pyrophosphate and FAD this is probably less significant. Nevertheless, the intracellular availability of a vitamin as a coenzyme to an enzyme which is dependent upon it, is believed to be a relevant index of vitamin status in the tissues.

The activity reached after the enzyme is incubated in vitro with the coenzyme, referred to as total or stimulated activity (S.A.), is believed to be representative of the potential activity if the coenzyme was not a limiting factor in vivo. It therefore measures not only the amount of holoenzyme functioning in vivo, but also the apoenzyme which would be undetected due to its inability to catalyze the reaction since it is uncomplexed. The result is the maximum potential activity. Certain aspects of coenzyme-apoenzyme relationships deserve mention in this context, although for practical purposes the above-mentioned description is probably adequate. It is conceivable that the amount of apoenzyme which is estimated by coenzyme activation is less than the total, since a proportion may be immeasurable by that means. One reason may be the inability of an enzyme to accommodate all of the coenzyme for which it has sites: the concept of catalytic cooperativity.[30] Alternatively, the stimulated activity could be the result of conditions which increase the binding capacity in vitro, but may not occur in vivo.

The enzyme that is used to study vitamin B_2 status is NADPH-dependent GR, one of three flavoproteins in the erythrocyte for which FAD is the cofactor. The reaction catalyzed by this enzyme was shown and discussed in Section I.B. The measurement of erythrocyte glutathione reductase (EGR) has been described on several occasions[26,31] and was suggested as an index of riboflavin nutrition in humans.[31,32]

The test was generally applied to erythrocytes, but whole blood can also be used. The reaction was usually monitored by the oxidation of NADPH to $NADP^+$, measured spectrophotometrically by virtue of the absorbance of the reduced form at 340 nm, or the reduced glutathione (GSH) was analyzed. A good correlation was observed[33] between these methods.

Since GR is inactive unless coupled with its coenzyme, it was believed that the activity detected in this manner reflected the amount of FAD available in vivo and therefore was an index of riboflavin status.[28,34] This could be altered in vivo by dietary B_2. The enzymatic method evaluated the metabolic utilization of dietary vitamin. The principles behind coenzyme activation of enzyme activity in vitro were also applied to GR inasmuch as the basal activity was measured as well as the degree to which the existing enzyme was unsaturated with FAD. The degree of in vitro stimulation of enzyme activity is dependent on the FAD saturation of the apoenzyme, which in turn is dependent on the availability of the vitamin. A number of relevant investigations has been carried out in the form of controlled experiments and in populations. From the S.A. and the basal activity (B.A.) was calculated the percentage stimulation or activity coefficient (A.C.), terms which have both been widely used in the literature.

According to a number of investigations with rats[31,35,36] and humans[26,28,32,34,37] GR activity is a useful and sensitive procedure for evaluating the status of both individual subjects and populations. The activity coefficient is believed to be a valid measurement of riboflavin status in animals[35,36] and in humans,[28] judging by depletion/repletion studies. The sensitive response of the enzyme to FAD in vitro was shown to be representative of other tissues.[35] In a recent biochemical evaluation of the test which was carried out in rats rendered acutely and chronically deficient, the relationship between the activity coefficient value and overall status in other tissues was tested. The assumption that these are closely related is valid, according to correlations between activity coefficient and other biochemical or physiological indexes of riboflavin status, including basal glutathione reductase activity, in a number of post-mortem tissues.

During riboflavin deficiency, it was found that erythrocyte glutathione reductase activity decreased with a concomitant increase in activity coefficient. When the intake of riboflavin could be controlled, the activity increased in response to the amount taken, an effect which leveled off when the activity coefficient was less than 1.25, indicating that this could be applied as a cut-off point to differentiate between inadequacy and sufficiency.[26] Reviewers of other studies[23,38] reached a similar conclusion. Subjects who had clinical evidence of a deficiency generally had a ratio of >1.25 but the reverse was not true. Several subjects with severe deficiency as judged by this criterion lacked clinical manifestations of a deficiency,[33] suggesting that the test was valuable in identifying a sub-clinical state. This view was confirmed in the application of the test to evaluate nutritional status in a population in which clinically recognizable vitamin B_2 deficiency was rare, therefore placing greater reliance on the biochemical test.[34] It was from investigations such as these that values of the activity coefficient have been proposed to aid the interpretation of the test.[23] An allowance of 0 to 20% stimulation (1.00 to 1.20 activity coefficient) provided a means of identifying the marginal subject and gave a reference range.[34] Even in the latter, the enzyme was not completely saturated with FAD, and therefore the in vitro addition of FAD commonly resulted in a small percentage increase.[32,37]

The data is usually expressed as activity coefficient or percentage stimulation, rather than enzymatic activities (basal and stimulated) since this offers several advantages. One is that the ratio appeared to be independent of age, sex, and laboratory method, allowing reference to a single guide, but nonetheless could only be used if the magnitude of the two component activities could be validly disregarded. Identical ratios could result from different combinations of these. One factor to be considered is the apoenzyme concentration. It is generally believed that the apoenzyme does not decrease.[27] In a depletion study[28] in which the daily allowance of 1.5 mg was reduced to 70 μg for 8 weeks, the ratio increased steadily from 1.10 to nearly 2.00 at the end of this period. This resulted in a considerable degree of unsaturation. A value of 2.00 means that 50% of the enzyme present is inactive due to the absence of FAD. Toward the end of the depletion period, there were signs that a limit had been reached. With repletion, it fell very rapidly to a normal level. That this could occur so quickly implied that the apoenzyme concentration had not been affected. In a smaller study[26] of this type of shorter duration, basal activity decreased and the percentage activation increased. However, the latter was low in some cases, indicating that stimulated activity, which represented total apoenzyme concentration, may also be affected.

There is evidence that the circumstances leading to the aforementioned situation may commonly exist. Pyridoxine deficiency tended to lower GR activity without the corresponding increase in FAD effect or activation coefficient and this was attributed to apoenzyme defect.[18] In pyridoxine-deficient animals it was noted that there was marked impairment of protein synthesis.[39] According to studies carried out by one group of investigators,[18] in riboflavin deficiency the drop in enzyme activity was generally associated with increased activity coefficient. The latter did not occur in pyridoxine-deficient animals, results which, when viewed together suggested apoenzyme loss when a shortage of the two vitamins coexisted. Therefore, the reliability of the activity coefficient as a measure of riboflavin status in populations in which deficiencies of vitamins B_2 and B_6 coexist, or in any other conditions in which apoglutathione reductase level is reduced, may be undermined.[40] With these considerations in mind, it was proposed that enzyme activity be taken into account as well as activity coefficient, and that riboflavin deficiency be suspected whenever the enzyme activity is low, even if the coefficient is normal.[40]

The measurement of GR activity and coenzyme activation has recently been applied to whole blood (BGR test).[41] This offered certain advantages, such as fewer manipulations. Very small volumes could be used, making it more convenient; furthermore, it could be automated. In an application of this procedure to a population of children, a normal range was 1.00 to 1.35. A criticism of the method was insensitivity, but it was felt that this may be outweighed by its convenience.[23]

According to reviews of the methods available to study B_2 status, the enzymatic approach was likely to be the method of choice.[23,33]

The selection of a particular criterion of vitamin B_2 status from several procedures available depends on an evaluation of many factors. These can be broadly grouped into practical or laboratory aspects, the characteristics of the subjects to be investigated, and the nature of the study to be undertaken. For the current study of vitamin B_2 status, it was deemed that the erythrocyte glutathione reductase test was the most appropriate procedure when considered in the light of other methods. One blood sample, which need not be fasting, was required, therefore it was a suitable procedure to use on outpatients. Notwithstanding the potential unreliability of results from urine samples, it was impractical to obtain these specimens under the circumstances. It was also necessary to use data which were not unduly influenced by recent intake of the vitamin, since the diet was not restricted, nor could it be monitored. Erythrocyte riboflavin had this advantage, however, it has been shown to be fairly resistant to controlled changes in dietary B_2. This made it difficult to detect subclinical changes occurring in a population of generally well-nourished women whose diet was essentially unaltered, but who were taking a drug (oral contraceptive agents) believed to alter vitamin B_2 status.

Experience with the erythrocyte glutathione reductase test in a number of laboratories has resulted in it being well recommended. In the light of the current study, the measurement of basal and stimulated activities and the calculation of activity coefficient should give a good indication of vitamin B_2 status. The methodology has been recently scrutinized[42] and the World Health Organization, as part of their program to standardize methods, has modified and adopted this technique.

II. MATERIALS AND METHODS

A. Experimental Design

The study was designed to compare a number of indexes of vitamin nutrition in a sample of control women and in oral contraceptive users and a cross-sectional approach was applied. In this investigation, results obtained from young women who had been using one of the types of oral contraceptive agents (OCA) currently marketed were compared with the results from a similar group of apparently healthy women who had never used them. There are a large number of formulations, but most people had been given a low dose brand, enabling its evaluation. Since the other types fell into several categories (based on dosage and constituents), the conclusions which could be drawn were limited. However, in this miscellaneous group resided all the high dose combinations. An attempt was also made to identify the importance of the length of time of oral contraception.

Extensive preliminary work preceded the implementation of these investigations. Among the practical aspects which required attention was the stability of the enzyme in fresh blood and also during the longer term in a laboratory. In these experiments replicate subspecimens were subjected to the assay technique over several periods of time.

It was also necessary to identify the reference range for the criterion of vitamin status chosen, before any conclusions could be drawn pertaining to the effect of the Pill on the female user community. The extent to which nutritional status could fluctuate was also relevant. Vitamin assays were therefore repeated on individuals on several different occasions.

B. Subject Selection

Female subjects were recruited randomly when they attended a family planning clinic in order to obtain an opinion about contraception. They were healthy women who had given their informed consent to participate. Each subject was interviewed initially for medical history and for relevant dietary information, and an outline of the study was explained. Volunteers living in the same region offered to provide blood specimens needed to clarify the laboratory aspects of the assay techniques and also to provide information about the expected levels of vitamins in members of the community.

The oral contraceptives used by the experimental subjects were not limited to one kind or even to a uniform amount of estrogen. However, approximately one half of the sample relied on a low dose type marketed under several names, which consisted of 30 μg ethynylestradiol and 150 μg d-NG. The remainder used a variety of formulations, the frequency and exact nature of which have been defined in the appropriate sections, but only a minority exceeded a 50 μg dose of estrogen or contained mestranol. The larger variety of progestogens currently available was also reflected in the sample. The duration of use by the participants ranged from a few months to several years.

C. Collection of Blood from Donors

Venous blood was withdrawn from an antecubital vein into a sterile, plastic syringe through a 21-gauge needle. After detaching the needle from the syringe, the blood was carefully delivered into a 10-mℓ plastic tube internally coated with heparin which prevented coagulation of the blood. The tube was gently inverted repeatedly in order to disperse the anticoagulant through the blood. It was important to ensure that hemolysis did not occur during the handling of the blood specimens. As this is sometimes caused by a sudden change in temperature, such as would occur when warm, freshly drawn blood is placed directly in crushed ice, the blood was allowed to cool naturally over a period of about 10 min. The tubes were then kept at 4°C in a refrigerator or, during field collections, in a container of crushed ice until further treatment. There was no evidence of clotting or hemolysis using the procedure adopted.

D. Separation of Erythrocytes and Plasma

Heparinized blood was centrifuged for 15 min at 5 km/s² at 4°C. The plasma was removed without disturbing the buffy coat, and was stored in a sealed plastic tube at −20°C. The buffy coat was then taken off and discarded. The remaining erythrocytes were washed once with an equal volume of a cold solution of 0.9% NaCl. Following centrifugation for another 15 min at 4°C, the supernatant was removed together with the top layer of erythrocytes to ensure that there was no contamination with white blood cells.

E. Storage of Erythrocytes

Enzymatic activity was measured in solutions of disrupted cells (hemolysates) prepared from the erythrocytes. It was not usually possible to carry out these assays immediately, but GR is less stable when stored as a hemolysate. A solution to this problem was to retain the enzymes within the intact erythrocytes until required, when the hemolysate could be prepared. This procedure enabled extended storage without appreciable loss of activity. As a single blood sample could serve as the source of other enzymes, in practice it was convenient to divide the packed erythrocytes into aliquots first, so that they could be stored independently and be accessible for future use. The procedure adopted was to remove an aliquot of 200-μℓ packed eythocytes immediately following the completion of the separation of erythrocytes and plasma. It was carried out using a pipette which was calibrated to deliver this volume accurately via disposable plastic pipette tips. The aliquot was delivered into the bottom of a small glass test tube (7.5 cm × 1.2 cm internal diameter), and the tube was sealed and stored frozen at −20°C. This technique was repeated to obtain the number of aliquots required, but usually at least six subspecimens were taken. The manipulation was carried out as quickly as possible, keeping the cells cold during this time. A fresh plastic pipette tip was used for each aliquot, maintaining a slow steady motion for the withdrawal and delivery of blood cells. These were removed from the center of the sample and excess blood on the outside of the tip was wiped off prior to delivery into the test tube.The subspecimens were stored for at least 1 day until shortly before the enzyme assay was carried out.

F. Preparation of Hemolysate

Before the procedure for obtaining a hemolysate from the frozen aliquot of erythrocytes could be applied, it was necessary for the sample to be maintained at −20°C for at least 24 hr. Shortly before the assay was to be carried out, the tube was removed from the refrigerator and allowed to thaw. The cells were lyzed by the addition of 3.8 mℓ phosphate buffer at pH 7.4 with thorough mixing. During this process, the tubes and solutions used were maintained at 4°C. Complete lysis was achieved in this manner and the cell coats were removed by centrifugation at 7 km/s² for 5 to 10 min. The supernatant solution was retained at 4°C, to be used as the source of erythrocyte glutathione reductase in the assay.

G. Assay of Erythrocyte Glutathione Reductase Activity

In the reaction catalyzed by GR (Section I.B.), NADPH was oxidized to NADP⁺. Since only the reduced form of NADP absorbs maximally at 340 nm, the reaction may be conveniently followed by spectrophotometrically measuring the rate of change in absorbance at 340 nm, this being a measure of the activity of the erythrocyte enzyme (EGR) present in the prepared hemolysate.

The experimental procedure was a modification of the method introduced by Beutler[32,43] and was the product of a WHO Task Force which studied vitamin methodology (Hall and Whitehead, unpublished data).

1. Materials

- Potassium phosphate buffer, 100 m*M*, pH 7.4.

- Glutathione (GSSG) in oxidized form (Sigma Chemical Company), 75 mM, containing sodium hydroxide at a final concentration of 7.92 mM (50 $\mu\ell$, 800 mM NaOH per 5 mℓ GSSG solution). This solution was prepared in a volume appropriate to the experiment, since it must be fresh, and was stored on ice.
- Sodium hydroxide (NaOH), 800 mM.
- Ethylene diamine tetraacetic acid (EDTA), potassium salt, 75 mM. This was freshly prepared each week and stored at 4°C.
- Sodium bicarbonate (NaHCO$_3$), 100 mM. This was prepared every 14 days and stored at 4°C during that time.
- Nicotinamide adenine dinucleotide phosphate (NADPH), tetrasodium salt, reduced form (Boehringer-Mannheim), 2 mM in 100 mM NaHCO$_3$. This solution was made up freshly each day shortly before use in a volume appropriate to requirements.
- Glutathione reductase (GR), in 3.2 M ammonium sulfate solution, specific activity approximately 120 units/mg (Boehringer-Mannheim). This was diluted with phosphate buffer to the level of activity required in the experiment.

2. Equipment

The spectrophotometer was a Unicam Model SP800, a double beam recording instrument with a cuvette holder which could be maintained at a constant temperature. This was achieved by circulating water from a thermostatically controlled water bath through the holder, allowing a reaction to proceed in the cuvette at a particular temperature and be monitored spectrophotometrically at the same time. An additional facility of the instrument was that the cuvette holder was controlled by an automatic cell changing attachment. There was provision for eight cells (four for the reference solution and four for the sample), so the contents of four cells could be measured simultaneously if desired. Each cell remained in the light path for 4 seconds and the magnitude of the absorbance was recorded on a chart which moved at 14.1 mm/min at a fixed wavelength.

3. Method

The reaction mixture consisted of:

100 mM Potassium phosphate buffer pH 7.4, 2.10 mℓ
75 mM GSSG, 100 $\mu\ell$
75 mM EDTA, 50 $\mu\ell$
Hemolysate, 100 $\mu\ell$
2 mM NADPH, 100 $\mu\ell$

in a total volume of 2.45 mℓ.

Buffered phosphate, GSSG and EDTA solutions were added to a small test tube (7.5 cm × 1.2 cm internal diameter) in the proportions shown. After a short pre-incubation time at 37°C, 100 $\mu\ell$ hemolysate was added and the incubation continued for 15.0 min in the shaking water bath at 37°C. Although this incubation period was not mandatory for the assay of the activity *per se*, the coenzyme activation (described in Section II.H.) was normally measured at the same time and this required a 15.0 min incubation of the enzyme with FAD. Uniform reaction conditions were therefore applied. The reaction began with the addition of 100 $\mu\ell$ NADPH solution. After mixing and swift transfer of the contents of the tube into a warmed spectrophotometer cuvette, the reaction continued at the same temperature in the spectrophotometer. This allowed the progress of the reaction to be continuously and directly monitored, since the consumption of one of the reactants (NADPH) resulted in a decrease in absorbance at the wavelength of 340 nm. At least 5 min was allowed to elapse before discontinuing the reaction. A maximum of four reactions could be measured simultaneously using all of the positions in the cuvette holder and automatically placing each cell in the light path every 34 seconds, but usually other manipulations precluded full use of this facility. A reference solution, 2.25 mℓ phosphate buffer and 100 $\mu\ell$ hemolysate, was used to adjust the baseline absorbance to zero and to use as a blank in the reference beam of the spectrophotometer. Distilled water would have sufficed, since the final absorbance after the addition of all of the solutions, was within the range of the instrument. The hemolysate, which served as the enzyme source for the assay, was retained in order to measure the concentration of hemoglobin (Section II.I.).

Purified enzyme, obtained commercially, was assayed in the same manner, except that 100 $\mu\ell$ of solution replaced the hemolysate. The change in absorbance was measured against a distilled water blank.

4. Estimation of ΔA/t (Time Rate of Change in Absorbance)

The rate of decrease in absorbance was proportional to the activity and therefore it was necessary to measure the gradient. A uniform rate of change throughout the duration of the reaction resulted in a linear slope, the magnitude of which was ascertained using two methods. The line of best fit was judged and drawn through the series of points provided by the recording. If each absorbance reading on the chart was noted, the corresponding time could be obtained with a computer program designed to calculate the gradient ($\Delta A/t$).

5. Calculation of Results

The decrease in absorbance at 340 nm corresponded to the oxidation of NADPH to NADP$^+$, and the rate of this conversion was dependent on the activity of the enzyme. Therefore, the change in absorbance per unit time was a measurement of the rate of the reaction. The higher the activity, the greater the $\Delta A/t$.

The unit of activity was mmol NADPH oxidized per hour per liter erythrocytes, i.e., the rate of NADPH oxidation per unit volume of erythrocytes (mmol/ℓ hr). This can be calculated from the formula:

$$\frac{\Delta A \times a \times b \times c \times d \times e}{t \times f \times g \times h}$$

where ΔA = change in absorbance during time t
a = volume of reaction mixture (ℓ)
b = conversion factor: mol to mmol
c = conversion factor to hr
d = total volume of hemolysate mℓ
e = total volume of RBC (mℓ)
f = molar extinction of NADPH at 340 nm
g = volume of hemolysate used in reaction mixture (mℓ)
h = volume of packed RBC (ℓ)
t = time for ΔA (min)

Therefore, mmol NADPH oxidized per hr/ℓ of erythrocytes

$$= \frac{\Delta A \times 2.45 \times 10^{-3} \times 10^3 \times 60 \times 4.0 \times 1.0}{t \times 6.22 \times 10^3 \times 0.1 \times 0.2 \times 10^{-3}}$$

$$= \frac{\Delta A}{t} \times 4726 \text{ mmol/}\ell \text{ hr}$$

This activity was termed **Basal Activity**.

H. Coenzyme Activation of Erythrocyte Glutathione Reductase

GR requires the cofactor FAD in order to function, and FAD is synthesized from the vitamin riboflavin. Therefore, the activity is dependent on the supply of riboflavin. The enzyme may not be saturated with respect to FAD at any one time, but a relatively constant proportion of FAD saturation is found in humans receiving adequate amounts of vitamin. An estimation of the amount of enzyme which is saturated with FAD was made from the basal activity, but in order to interpret this result further it was necessary to determine what proportion of the total cellular pool of enzyme was saturated with FAD. Since the apoenzyme was unable to catalyze the reaction, it could not be measured by the procedure described earlier. However, this problem may be overcome by incubating the enzyme in the presence of FAD, supplied in vitro in excess, thereby rendering apoenzyme catalytically active. When enzymatic activity is remeasured by the same assay procedure, this should be equal to or greater than the original activity. From these two results could be calculated the proportion of the enzyme which was saturated with FAD.

Coenzyme activation was carried out by a modification (Hall and Whitehead, unpublished data) of the method published by Beutler,[32,43] commissioned by WHO.

1. Materials

With the addition of FAD, the solutions were those described in Section II.G: FAD, disodium salt (Sigma Chemical Company), 300 μM. The solution, which was protected from the light, was prepared shortly before it was required for the assay and was retained in an ice bath.

2. Equipment

Spectrophotometer, as detailed in Section II.G. When coenzyme activation of enzyme was measured, it was usually accompanied by the assay of basal activity. Uniform conditions could be applied by transferring the two reaction mixtures to cuvettes and setting the spectrophotometer to read them consecutively to provide the absorbance readings.

3. Method

The procedure described in Section II.G. to measure GR was used, except that 100 $\mu\ell$ FAD was included in the reaction mixture of 2.45 mℓ. The two assays were carried out in parallel, care being taken to protect the FAD solution from the light. Phosphate buffer containing GSSG, FAD, and EDTA was warmed to 37°C, then 100 $\mu\ell$ hemolysate was added and the incubation continued for 15.0 min with shaking. NADPH solution (100 $\mu\ell$) initiated the reaction and the contents of the tube were mixed and rapidly transferred to a warm cuvette, so that the progress of the reaction could be recorded in the spectrophotometer where the temperature was maintained at 37°C. The reaction mixture that contained FAD could be monitored at the same time as the one which lacked FAD by using the cell holder which could automatically move each cell into the light path. Each absorbance was measured every 25 seconds for 4 seconds at a time. One blank solution, made up from 4.5 mℓ phosphate buffer and 200 $\mu\ell$ hemolysate and divided in half could be used for the two solutions.

Coenzyme activation of the purified enzyme (GR) was not normally attempted since the enzyme was supplied already saturated with coenzyme.

4. Estimation of ΔA/t (Time Rate of Change in Absorbance)

The time rate of change of absorbance (ΔA/t) could be measured in the same way as described in Section II.G. By measuring the activity in the presence and absence of FAD simultaneously, the slopes could be calculated from one graph. Generally, the slope was greater if the enzyme had been incubated in the presence of FAD since this caused an increase in the activity.

5. Calculation of Results

The activity in the presence of added coenzyme was expressed as mmol NADPH oxidized per hour per liter erythrocytes, i.e., the rate of NADPH oxidation per unit volume of erythrocytes (mmol/ℓ hr). This was calculated from the formula in Section II.G.

This activity was termed **Stimulated Activity**.

I. Chemical Analysis of Hemoglobin (Hb)

Hemoglobin concentration in a disrupted cell preparation (hemolysate) was determined by the cyanmethemoglobin method.[44] Blood was diluted with a potassium ferricyanide-cyanide solution which oxidized hemoglobin to methemoglobin. The latter was converted to cyanmethemoglobin which absorbed maximally at 540 nm. The technique determined the sum of oxyhemoglobin, hemoglobin, methemoglobin, and carboxyhemoglobin.

J. Statistical Analysis

The analysis of the data relied on several methods, and these have been described in Appendix A. The application of these statistics to the data is located within Section III.

III. RESULTS

The results obtained during this study were classified into three main areas and have been presented below. In the first section, the laboratory aspects were considered. Vitamin B_2 status, as determined by the erythrocyte glutathione reductase test, was then defined in a sample of women chosen to serve as a control group. This information was a foundation on which to base the results from another sample of women, the oral contraceptive users.

A. Assessment of Vitamin B_2 Status by the Glutathione Reductase Test (Laboratory Aspects)

1. Measurement of $\Delta A/t$ (Time Rate of Change in Absorbance)

During the reaction catalyzed by GR, NADPH was oxidized to $NADP^+$, with concomitant reduction of glutathione. Therefore, the enzymatic activity was proportional to the consumption of NADPH as measured by the rate at which the absorbance decreased. The record of this reaction was graphical. For this assay, the visual estimate of "line of best fit" could be achieved without undue difficulty because the reaction rate was constant. There was good agreement in a comparison between the visually fitted gradient and that calculated from the series of points. The error incurred in finding the slope was estimated by drawing several lines through the same data points. Although the variation depended on the data used, the average was ± 3.6 mmol/ℓ hr. Since this was based on a number of possible lines, in practice the error was less when adhering to a consistent technique. A different error was introduced when the absorbance value was read from the chart recording and this was estimated to be of the order of ± 0.005, which was equivalent to a maximum error of about 5 mmol/ℓ hr (± 2.5). Therefore, the error in finding the correct slope and in reading the absorbance axis could theoretically amount to ± 6 mmol/ℓ hr, but in practice the two were probably not mutually exclusive. The other main source of error was in the experimental technique. The combined effect of these errors was reflected in the reproducibility of the results, an analysis of which is presented in Section III.A.5.

2. Expression of Results

A number of different expressions can be applied to the data obtained from enzyme assays. During the analysis of the erythrocyte glutathione reductase results, the following expressions were used:

1. Activity of EGR, i.e., time rate of NADPH oxidation per unit volume of erythrocytes (mmol/ℓ hr).
2. Activity coefficient, i.e., stimulated activity/basal activity.
3. Percentage saturation of enzyme with cofactor, i.e., (basal activity/stimulated activity) $\times$ 100.

3. Stability of Glutathione Reductase in Blood Samples

It was believed that the enzyme was stable for 8 hr at 4°C and even at 37°C. However, the maximum time which was allowed to elapse between drawing of blood and laboratory processing was 3 hr and the activity was not affected appreciably by temporary storage at 4°C. It was found that when the blood was kept 13 hr at 4°C before processing, both the basal and stimulated activities were lower, but the ratio of these (activity coefficient) was the same as an identical sample which had been treated in the usual way.

4. Erythrocyte Subspecimens

Numerous subspecimens of intact erythrocytes of 200 $\mu\ell$ each were dispensed from a blood sample into tubes for storage until the enzyme assay could be performed. The accuracy with which this could be done was measured in several ways:

1. **Weight:** The aliquots of blood required for all of the vitamins were dispensed at the same time. The mean weight was 0.179 $\pm$ 0.009 g (S.D.).
2. **Hemoglobin concentration:** The solution which was assayed for enzymatic activity was obtained by diluting a thawed specimen of cells from 0.2 to 4.0 mℓ by adding phosphate buffer. The resultant solution was also subjected to hemoglobin analysis and therefore it was possible to estimate the similarity between duplicate subspecimens on the basis of the concentration. Using the results from a number of paired hemolysates it was estimated that the variation was 8.4% and the method variation was 5.4% (coefficient of variation).
3. **Enzyme activity:** The delivery of a particular volume was also measured using the GR activity of replicate subspecimens of erythrocytes which had been hemolyzed to give an enzyme solution. A comparison of the duplicates led to the conclusion that the coefficient of variation was 4.3%.

5. Assay Variation in Glutathione Reductase Activity

Several sources of error could be identified in the assay of GR activity, one being the abstraction of results directly from the instrument recording (discussed in Section III.A.1.). The manipulations required to produce the reaction mixture for this quantitation were also potential sources of error. Although the identification of individual steps which could contribute variability to the results was relevant, particularly if it became necessary to make improvements, it was the cumulative effect on the final result which was considered. The overall reproducibility of the method for measuring activity would be a reflection of the experimental technique.

The measurement of activity in a blood sample was replicated, usually in duplicate or triplicate. Within-run precision was therefore estimated by analyzing routine unknowns in duplicate and using the paired results to calculate the precision according to the method described in Appendix A.

Some data was obtained when the assay was performed in duplicate on each hemolysate taken from a group of healthy volunteers. The coefficient of variation was estimated to be 2.7% for 164 determinations ranging in activity from about 73 to 170 mmol/ℓ hr, and when these results were analyzed as three ranges of activity (low, medium, and high), the coefficient of variation ranged from 2.2 to 4.3%.

The reproducibility which existed between runs was relevant since usually no more than ten specimens could be managed in each run, and blood was submitted for analysis at regular intervals. Replicate subspecimens were stored frozen and each one of these was analyzed on a separate occasion. This was done the same day or on consecutive days in order to minimize any effect storage time may have on enzyme activity. An hemolysate was not retained for remeasurement in a succeeding run because of the doubtful stability of the enzyme in this form. The paired results for 22 blood samples submitted formed the basis of the between-run variation of 3.6% (coefficient of variation) for a group of samples with mean activity 129.4 ± 16.3 mmol/ℓ hr (S.D.). Two ranges could be identified. It was found that this variation was greater than that occurring within-run, a result which would be expected.

6. Storage of Blood Specimens

The storage of blood specimens was not a particular problem, since assays were usually carried out within a few days of collection. However, it was useful to know what effect time could have on the activity of GR. Maintenance of cells intact until shortly before an assay was an advantage over earlier versions of the method, in which cells were lysed and the hemolysate was stored until required.

Replicate subspecimens were thawed 1, 2, and 6 days after blood was drawn, and there appeared to be no deterioration in activity. In another experiment, after 4 and 22 days, specimens from 12 volunteers did not exceed the expected between-run variation which had been found for the assay. Any differences in basal and stimulated activities were parallel, and therefore the ratio was constant. Over the longer term, 1.5 and 3 months after the blood had been taken, storage had an effect on the results, causing decreases, and during this time an insoluble material appeared in the solution which could be removed by centrifugation.

7. Control of Laboratory Error

The inclusion of a standard in routine runs may indicate whether the technique of the method is satisfactory. The provision of a suitable standard for the assay of GR is currently being arranged on behalf of WHO in order to compare the results from different laboratories using identical experimental procedures. It is anticipated that material of unknown enzymatic activity would be submitted to different laboratories for analysis and the results could be compared. It would be an advantage if this material was stable and could be used regularly as a standard. However, participation in this program is not yet possible, and therefore other methods were used to check laboratory error:

1.   Erythrocytes were stored frozen as a large number of subspecimens, one of which could be thawed and measured each time the test was performed. This was not very satisfactory since they could only be stored for a relatively short length of time.
2.   Purified GR was available commercially and could be assayed for activity at the same time as the unknown hemolysates. The specific activity was considerably greater than that in the erythrocytes and a dilution was made so that the activity was of the same order as that encountered in the unknowns. The resultant solution was divided into aliquots and frozen at −20°C until required. This enzyme preparation was already saturated with FAD and therefore it was not necessary to carry out the assay in the presence of added FAD.

8. Comments

Personal experience showed that the assay could be reliable. While not difficult, it did require care and attention to details, and also had certain disadvantages: (1) it was very labor intensive at nearly every stage of the procedure; (2) in order to make this test available routinely and economically, it would be necessary to use an instrument that had a better form of data output; (3) the technique was inconvenient for small numbers of specimens which were submitted irregularly due to the amount of preparation required irrespective of the number; and (4) factors such as the stability of solutions limited the quantity which could be run.

B. Application of the Erythrocyte Glutathione Reductase Assay to the Evaluation of Riboflavin Nutritional Status in Normal Female Subjects

The hypothesis that women who are using an oral contraceptive incur a change in vitamin B_2 status was tested with reference to a control group. Ranges of these values were therefore required for the criterion which was employed, erythrocyte glutatione reductase (EGR). The two measurements, basal activity and stimulated activity, gave rise to two more variables, activity coefficient and percentage saturation (P.S.), and these all formed the basis of the assessment of vitamin B_2 status.

1. Probability Distribution of Results

Erythrocyte glutathione reductase activity and activation with FAD were measured in women who were taking estrogen-progestogen preparations, but no conclusions could be drawn from these results unless reference was made to those from another group of women, the control sample, who did not take oral contraceptives. Statistical methods were required to analyze this data, and to ensure that the application was appropriate, it was necessary to determine the probability distributions.

Figure 3 shows the histograms of four indexes of vitamin B_2 status in the control sample, representing the basal activities and stimulated activities, and the ratio's activity coefficient and percentage saturation. The type of frequency distribution in each case was sought by two different calculations, the details of which are in Appendix A. Both methods applied the concepts of the standardized normal distribution and the χ^2 test. One of these was based on a comparison of the observed and expected frequency of observations within each class interval. However, total reliance was not placed on this method because the outcome can depend on the choice of class interval. This was eliminated by designating areas of equal probability within the standardized normal distribution and comparing the expected and observed frequencies within each area. It was concluded that each of the four variables conformed to a normal probability distribution. However, other distributions were also explored since the ratio of two variables which are from normally distributed populations does not usually result in a quotient which is similarly distributed (R. Gollan, personal communication).* These calculations showed that the logarithmic function of the basal activity ($\log_{10} BA$) was also normally distributed, meaning that the original values conformed to logarithmic normal, i.e., a curve skewed to the right. Stimulated activity was just within the limits required for a normal distribution, but $\log_{10} SA$ was not. Despite this, the application of accepted statistical tests for normality justified the view that basal activity, stimulated activity, activity coefficient, and percentage stimulation were normally distributed variables, and therefore parametric statistical methods were applied.

2. Reference Ranges

Young women who were not taking additional vitamins or other medications were the source of the results shown in Table 1. Complete saturation (100%) is equivalent to an

---

* The ratio of two standard normal random variables has a Cauchy distribution, having the form

$$f(u) = \frac{1}{\pi (1 + u^2)}, \quad -\infty < u < \infty.$$

This distribution is symmetrical about mean zero.

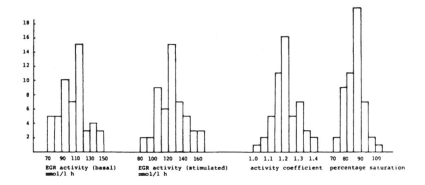

FIGURE 3.   Frequency distribution of glutathione reductase activity and activation.

Table 1
REFERENCE RANGES FOR
ERYTHROCYTE GLUTATHIONE
REDUCTASE ASSAY

	Arithmetic mean and standard error
EGR activity (basal) mmol/ℓ hr	107.3 ± 2.6
EGR activity (stimulated) mmol/ℓ hr	125.1 ± 2.7
Activity coefficient	1.17 ± 0.01
Percentage saturation	85.6 ± 0.9
Number	52

activity coefficient of 1.00. In this sample, the mean was estimated to be 1.17 and according to the standard deviation (0.08), 95% of the values lay between 1.01 and 1.34. The percent saturation ranged from 73 to 98.

The expected level of activity depends on the particular technique, but a comparison can be made between the results obtained from this sample of people and those obtained from other populations in terms of the ratios, since methodological variations can then be disregarded. Ranges of activity coefficient for different states of vitamin B_2 nutriture have been derived from many different types of study and the guidelines suggested by Sauberlich are shown in Figure 4. The estimated mean of 1.17 is within the range of acceptable vitamin B_2 status and therein lay 67.3% of the sample (Table 2). The remainder (32.7%) were classified as low according to this criterion, but none were deficient. It has also been proposed that up to 1.25 should be considered acceptable and as this represents approximately 1 S.D. above the mean, it may be a more reasonable limit to apply to the population from which this sample was drawn, with deficiency exceeding 1.34 (2 S.D.).

3. Relationship Between Basal and Stimulated Activity

In Table 1 it can be seen that the mean stimulated activity was 1.17 times greater than the mean basal activity. This figure, which is referred to as an activity coefficient, is a measure of the degree of enzyme saturation and has gained acceptance as an index of vitamin B_2 status. This is based on the premise that during experimentally induced riboflavin deficiency in an individual, the basal activity varies according to the concentration of FAD, and as stimulated activity remains relatively constant, a decrease in basal activity would be accompanied by an increase in activity coefficient. There are advantages in using this term

```
ERYTHROCYTE
GLUTATHIONE REDUCTASE    =      EGR   +   FAD
(EGR) INDEX                     EGR   -   FAD


ACCEPTABLE                      1.00  -  1.19


LOW                             1.20  -  1.39


DEFICIENT                        >  1.40
```

FIGURE 4.   Guidelines for the interpretation of erythrocyte glutathione reductase activity coefficient.[23]

Table 2
VITAMIN B$_2$ STATUS:
CLASSIFICATION OF REFERENCE
SAMPLE

Activity coefficient

1.00—1.19	1.20—1.39	≥1.40
(acceptable)	(low)	(deficient)
67.3%	32.7%	0%

rather than basal activity when referring to vitamin B$_2$ status, and most of the published results are quoted in this form. The evaluation of oral contraceptives (to be discussed later) is based on the same test, making it necessary to ensure that appropriate indexes are used. Therefore, the reliance which could be placed on the ratio as a reflection of the basal activity and as an index of vitamin B$_2$ status in a sample, rather than in an individual, was tested by studying the relationship between the activity coefficient and the basal activity using the same results which are grouped in Table 1. It was anticipated that this relationship would also be of an inverse nature, such that a lower level of activity would be associated with a higher ratio and vice versa.

Figure 5 shows that there was an inverse relationship between basal activity and activity coefficient (r = -0.47).* Although the correlation coefficient was statistically significant, ($p < 0.0005$), the value of r^2 was more relevant, since this figure indicates whether the relationship is useful in a predictive sense. In this case, r^2 was 0.22, implying that 22% of the variation in y can be attributed to variation in x. Although there was a statistical relationship between the variables, it was limited in a predictive sense. At the upper end of the activity range the activity coefficients were uniformly low and, in general, the lower activities were associated with the highest activity coefficients. However, there were two subjects in whom this was accompanied by an acceptable ratio. If the enzymatic activity had been disregarded, by referring only to activity coefficient, these would not have been noticed. This end of the range is particularly important because it represents a tendency toward a state of depletion. In this sample, the latter would be expected to occur at low frequency, and therefore it was difficult to define this area. Reference has been made to some of the circumstances which may lead to both lowered basal and stimulated activities, resulting in a normal ratio. When this occurs, misleading conclusions can be drawn con-

---

*   The original presentation of data in this manner revealed an outlier which, when included, had a disproportionate effect on the correlation coefficient. The outcome of the correlation and regression analyses was therefore quoted without this result, believed to be due to supplements.

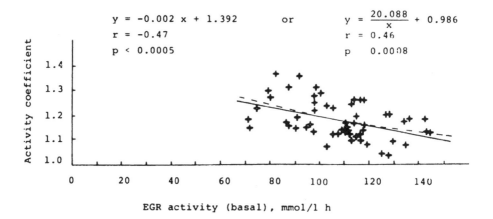

FIGURE 5.   Relationship between glutathione reductase activity (basal) and activity coefficient.

cerning the vitamin B_2 status of an individual. Therefore, in the present study, both basal activity and activity coefficient were taken into account.

A visual examination of Figure 5 indicated that several other regression curves could be fitted to the same observations. These were of the types described by Equations 2, 3, and 4:

$$y = -ax + b \tag{1}$$

$$y = \frac{a}{x} + b \tag{2}$$

$$y = -ax^2 + bx + c \tag{3}$$

$$y = be^{-ax} \tag{4}$$

where   y =   activity coefficient
         x =   basal activity
         a,b,c =   constants

The aforementioned correlation coefficient referred to the relationship described by Equation 1 for which the corresponding regression line was

$$AC = -0.002\ BA + 1.392$$

However, since several relationships were possible, other correlation coefficients were tested. Transformation of Equations 2 and 4 into a linear form $(Y = aX + b)$ by using functions of x and y and plotting the appropriate functions (Y against X), enabled the linear correlation coefficients to be calculated. The functions which applied to Equations 2 and 4 were $Y = y$, $X = 1/x$ and $Y = \ln y$, $X = x$, respectively. For Equation 3, a multiple correlation coefficient was calculated.

There was very little difference between the correlation coefficients and therefore, the strength of the relationship was described by the r value, which was about 0.48. Although a number of different models could be fitted, either a straight line or a curve, in view of the relationship between basal activity and stimulated activity (to be discussed shortly), the more appropriate models were

$$Y = -aX + b \quad \text{where } Y = AC, X = BA$$

$$Y = aX + b \quad\ \ \text{where } Y = AC, X = 1/BA$$

Therefore, the curve of best fit for Figure 5 was either

$$AC = -0.002 \text{ BA} + 1.392 \qquad \text{(a straight line)}$$

or

$$AC = \frac{20.088}{\text{BA}} + 0.986 \qquad \text{(a hyperbola)}$$

This was valid only over the range represented by data. Extrapolation of a curve would be particularly unwise at the left-hand side of the graph despite the upward tendency. Physiologically, the minimum activity coefficient is 1.00, whereas the maximum value is uncertain. During enforced depletion experiments, this may exceed 2.00, but may not do so under natural circumstances. During experiments where an individual was deprived of riboflavin and monitored by the increase in activity coefficient, the inverse relationship between this and basal activity would be described by Equation 2 since, by definition, AC = SA/BA. The aforementioned analysis was carried out in order to test whether results from many people, measured once and representing a range of activities, could be considered analogous to those obtained from one person measured many times over a range of activities which were produced by depletion. Although the range in the reference sample of volunteers was limited, in principle the same relationship (Equation 2) should apply.

Basal activity and activity coefficient were related, and therefore, since AC = SA/BA, it followed that there was a relationship between basal and stimulated activities. According to the curves calculated for Figure 5, this could be estimated by either

$$SA = -0.002 \text{ BA}^2 + 1.392 \text{ BA}$$

or

$$SA = 0.986 \text{ BA} + 20.088$$

Figure 6 is a scatter diagram of basal activity plotted against stimulated activity, for which the correlation coefficient was 0.93 and the regression line was

$$SA = 0.96 \text{ BA} + 22.35$$

As the data appeared to follow a slight curve, a parabola might also be appropriate:

$$SA = -0.002 \text{ BA}^2 + 0.987 \text{ BA} + 21.30$$

The multiple correlation coefficient was also 0.93, and the resultant slight curve and the straight line were superimposed and equally applicable mathematically. On the basis of earlier arguments, the parabola was less likely. The value of r^2 was high (0.86) and therefore the regression curve could be used as a prediction model for stimulated activity, given basal activity and vice versa. In practice, these values are not used very often. According to Figure 5 the activity coefficient of 1.19, above which vitamin B_2 status may be described as low, corresponded to basal activity of 100 mmol/ℓ hr and this was accompanied by stimulated activity of 119 mmol/ℓ hr (Figure 6). Therefore, a general guide to lowered vitamin B_2 status might be BA < 100, SA < 119, AC > 1.19.

The ideal vitamin B_2 status could be envisaged as that in which all of the GR is saturated with FAD. That is, the concentration of FAD in vivo is such that all of the enzyme is catalytic and therefore the levels of basal and stimulated activity are similar, resulting in a ratio of 1.00. Figure 6 shows that this situation did not exist in any subjects and that there was no interaction between the line represented by optimum activity coefficient and the

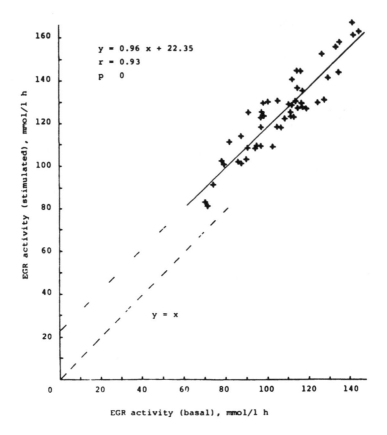

FIGURE 6.   Relationship between glutathione reductase activities measured before
and after stimulation with FAD.

regression line shown. The wide range of total activity in the sample represented differences
in the total amount of the enzyme. In each individual, this is probably maintained relatively
constant, with altered B_2 status being reflected in variable basal activity. However, in a
group, similar degrees of FAD saturation could be attained over a range of total enzyme
concentration. In Figure 6, as stimulated and basal activity decreased, the ratio of stimulated
to basal activity would increase, with the values of BA of about 100 and SA of about 119
possibly being critical values. However, there was no correlation between the activity coefficient and stimulated activity and therefore the latter could not be regarded as an indicator
of vitamin B_2 status.

The percentage saturation was also proportional to basal activity, an outcome which was
expected since it is the reciprocal of the activity coefficient. Figure 7 illustrates this relationship. As basal activity increased, the percentage saturation was also greater, but the
latter varied over a relatively smaller range (74 to 96%) for a corresponding range of 70 to
145 mmol/ℓ hr activity. It appeared that the proportion of the enzyme which was saturated
did not fall below about 74% under these stable conditions even when the basal activity was
low. However, in human volunteers under different circumstances, much lower percentage
saturation has been encountered in experimentally induced states of vitamin B_2 depletion.
These usually took place over a relatively short period of time.

4. Normal Variation

The assessment of vitamin B_2 nutritional status can be made according to a number of
criteria, but one of the reasons for adopting the enzymatic method (erythrocyte glutathione

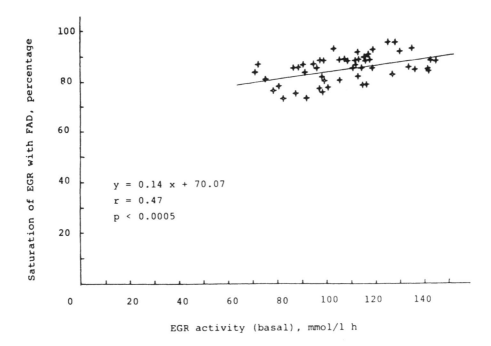

FIGURE 7. Relationship between glutathione reductase activity (basal) and percentage saturation.

reductase activity and activation) was because it is not susceptible to short-term variations in FAD concentration, and as such should give a good impression of the current vitamin B_2 status of an individual.

The week-to-week variation in GR activity and activation was evaluated in a series of experiments which was carried out on a group of 12 volunteers aged 18 to 25 years who donated blood at regular intervals. Apart from one male, the subjects were female, none of whom was taking a form of oral contraceptive pill. The average age of the females was 19.7 ± 1.3 years (S.D.). Blood was drawn on five occasions (days 0, 7, 14, 21, and 31), referred to as samples A, B, C, D, and E, respectively. The results from this series of measurements provided the basis of an assessment of the variation in vitamin B_2 status of the donors (using the criterion of GR) by considering them as a group. Table 3 shows an example of the results obtained (activity coefficient only). It was expected that there would be differences among the means for the subjects. This is illustrated in Table 4, which lists the means of five results for basal activity, stimulated activity, and activity coefficient, using 12 different volunteers. In order to evaluate individual variability and to find out whether the passage of time was a factor which altered any of these indexes, a two-way analysis of variance technique (see Appendix A) was applied. The five means of activity or activity coefficient for the group (Table 5) could not be compared for the five different times using one-way analysis of variance, because there were different characteristics for each individual. The subject to subject variability could, however be filtered out. The effect of one variable, i.e., the subject, on the criterion variable (basal activity, stimulated activity, or activity coefficient) could be eliminated in two-way analysis of variance, making it possible to evaluate the effect of the other variable, time (or sample). Table 3 depicts an example of a randomized block design used to solve the problem.

The suspected differences among the 12 subject means (Table 4) were substantiated by the analysis for both of the activities and also for the activity coefficient. It was concluded that there were no differences among the five means representing the multiple samples when activity coefficients were used, but it was also found that when either of the activities was

Table 3
EGR ACTIVITY COEFFICIENTS BY SUBJECT AND SAMPLE

(Two-way analysis of variance)

Sample[a]	Subject											
	1	2	3	4	5	6	7	8	9	10	11	12
A	1.17	1.06	1.04	1.49	1.16	1.17	1.08	1.23	1.11	1.11	1.13	1.08
B	1.21	1.09	1.04	1.35	1.06	1.29	1.18	1.36	1.25	1.22	1.17	1.21
C	1.17	1.11	1.11	1.39	1.12	1.24	1.13	1.23	1.20	1.04	1.19	1.14
D	1.10	1.10	1.04	1.39	1.00	1.25	1.16	1.21	1.34	1.18	1.06	1.13
E	1.07	1.18	1.11	1.33	1.05	1.27	1.12	1.15	1.25	1.25	1.10	1.14

[a]   0-, 7-, 14-, 21-, and 31-day blood samples.

Table 4
SUMMARY OF THE SUBJECT
MEANS FROM MULTIPLE SAMPLES

	Glutathione reductase		
Subject	Basal activity[a]	Stimulated activity[a]	Activity coefficient
1	100.7	114.4	1.14
2	130.9	144.6	1.11
3	115.6	121.1	1.07
4	109.7	151.5	1.39
5	111.1	119.3	1.08
6	108.7	134.6	1.24
7	137.7	155.7	1.13
8	104.8	129.1	1.24
9	104.7	128.5	1.23
10	111.2	128.2	1.16
11	126.9	143.2	1.13
12	133.5	151.9	1.14

Note: Each number represents mean calculated from 5
values (n = 5).

[a]   mmol/ℓ hr.

considered, there were differences among these five. In this type of analysis, a question to be considered was whether any interaction existed between the two factors. In other words, did a difference in mean activity remain constant for all subjects? This was addressed using a $2 \times 12 \times 5$ factorial design (two results for each of 12 subjects on 5 separate occasions), whereby it was concluded that there was no interaction, meaning that any difference in activity (basal and stimulated) between two or more occasions was constant for all of the volunteers.

Normal variation of vitamin B_2 status would probably take a different form in each person, and therefore the aforementioned result, indicating a similarity in response, suggested that the variability did not originate from the individuals concerned. The difference could be the results of consistent errors such as those which contribute to variation between runs. If this applied to both basal and stimulated activity, then the difference would be minimized in the

Table 5
SUMMARY OF THE SAMPLE MEANS FROM A SUBJECT GROUP

(Tukey's W procedure)

	Sample				
Glutathione reductase	**A**	**B**	**C**	**D**	**E**
Basal activity[a]	116.4	121.8	111.0	113.1	119.3
Stimulated activity[a]	132.1	145.1	129.0	130.9	138.7
Activity coefficient	1.15	1.20	1.17	1.16	1.17

Note: Each number represents mean calculated from 12 values (n = 12).

[a] mmolℓ hr.

ratio. The outcome of the analysis using activity coefficients supported this view, since there were no differences among these means for the five samples.

When the means were ranked in order of magnitude, (C < D < A < E < B) basal and stimulated activities were similar. Differences among means had been found using the above-mentioned methods, but it was not known which levels differed from the rest, nor the magnitude of the difference. A multiple comparison method, Tukey's W procedure (Appendix A) showed that

$$
\begin{array}{ll}
B > C,D & \text{Basal activity} \\
E > C & \\
B > A,C,D & \text{Stimulated activity} \\
E > C,D &
\end{array}
$$

The similarity in pattern could explain why no changes were seen for the ratio. Table 5 shows the numerical results. The means were judged to be significantly different from each other if they differed by an amount calculated from the procedure described.

The view that the source of variability was mainly a function of the experimental technique rather than normal individual variation in GR could be considered in the light of the laboratory aspects discussed in Section III.A. One of these was the between-run variation in GR activity. The differences which were found were within the limits of variability from between-run imprecision. Although this accounted for the variations, individual fluctuations were not entirely discounted, but they were likely to be small. Therefore it appeared that vitamin B_2 status (as measured by the GR test) monitored over a period of 4.5 weeks did not change appreciably.

C. Application of the Erythrocyte Glutathione Reductase Assay to the Evaluation of Riboflavin Nutritional Status in Females Taking Oral Contraceptive Agents: Cross-Sectional Study

The belief that women who are using oral contraceptives incur an adverse effect on their vitamin B_2 status was tested by comparing a sample of these people with a control sample. The approach to the problem required two different samples to be drawn from the same population (cross-sectional study). The GR assay in erythrocytes was the criterion used to describe vitamin B_2 status, and this has already been defined with respect to the four indexes: basal activity, stimulated or total activity, activity coefficient, and percentage saturation, in a group of women who exhibited the ranges usually encountered (Section III.B.). There are advantages in lowering the dose of steroid components in the Pill and now that there is substantial patronage of those containing 30 μg ethynylestradiol in combination with 150 μg d-NG in favor of other higher doses, it was determined whether these merits apply to vitamin status.

The nature of the prescriptions for the study sample is shown in Table 6. Fifty-four percent (54%) of these were of the one type, the low dose Pill. Many other combinations were represented, but most of these contained 50 μg estrogen (usually ethynylestradiol) in combination with miscellaneous progestogens (high dose Pill). The estrogen component is believed to be responsible for many of the clinical and biochemical side effects, and these are often dependent on the dose level. These two main subgroups were therefore compared, but there were insufficient numbers within each of the high dose Pill combinations to discriminate further.

The data obtained from the entire sample of these females, together with that from the control sample were analyzed and the results were summarized as the mean and standard error (Table 7). After analysis by the ''F'' and ''z'' test (Appendix A), certain differences emerged between the two groups. The mean basal activity, stimulated activity, and percentage saturation were all lower in the treated group, and in keeping with these indications of an adverse effect on vitamin B_2 status, the mean of the activity coefficient was greater ($p <$ 0.025) as was the variance. In terms of the guidelines for classifying states of vitamin B_2 nutriture based on the activity coefficient (Figure 4) the mean was within the acceptable range in the controls, but in the low range for the other group. A more detailed picture can be obtained from the first two lines of Table 8 which revealed that 12.7% of oral contraceptive users could be classified as being deficient and 46% were in the low or deficient category as compared with 33% of the controls (none of these were deficient). Application of the χ^2 test (Appendix A) confirmed that the profile of vitamin B_2 status in the oral contraceptive sample was not the same as that which had been observed in the reference sample ($p <$ 0.005) and that their status was worse. It has been mentioned previously that limits other than 1.20 have been suggested as the boundary below which vitamin B_2 status may be considered as satisfactory, but this does not invalidate the conclusion.

Table 9 represents a comparison within the sample of oral contraceptive users, which was made in order to see whether the deterioration in B_2 status could be attributed to the type of pill. Two major types of these were identified, based on the estrogen dose. Those containing 30 μg ethynylestradiol were combined with 150 μg d-NG and these formed one subgroup, low dose; those exceeding 30 μg of an estrogen were in combination with a variety of progestogen types at higher doses, hence the designation high dose. There was no significant difference between the two subgroups for any of the four variables, indicating that the adverse effect already noted could not be attributed to one of the categories. Each subgroup was also compared with the control group, as shown in Table 9 using one-way analysis of variance. This test demonstrated that at least two of the means of the basal activities were different ($p <$ 0.066), but those of the other variables were similar ($p >$ 0.1). It was deduced that both of the oral contraceptive subgroups were less than the controls in terms of the basal activity, and the low-dose subgroup was least when the means were ranked. These conclusions were also reflected in Table 8 which shows that 17.7% of the subjects taking the lower dose Pill were affected sufficiently to be classified as deficient, as compared with 6.9% of the higher dose subgroup. In other words, three fourths of the deficient subjects were taking the lower dose Pill. According to the χ^2 test, the frequency noted in the control sample was not reproduced in the high-dose subgroup ($p <$ 0.025) and the difference was even greater within the low-dose subgroup ($p <$ 0.005).

The outcome of these group analyses was that it conformed to the view that oral contraceptives exerted an adverse effect on B_2 status when GR was the criterion. Evidence for this came from the lowering of enzymatic activities and less saturation of enzyme with its cofactor. This was reflected in the higher activity coefficient. According to the latter, all of

Table 6
TYPES OF COMBINED ORAL CONTRACEPTIVE PREPARATIONS AND NUMBER OF EACH TYPE USED BY THE STUDY SAMPLE

| | Estrogen (µg) | | | | |
| | Ethynylestradiol | | | Mestranol | |
Progestogen	30	35	50	50	75
d-NG (150 µg)	34				
d-NG (250 µg)			7[a]		
Norethisterone (500 µg)		3			
Norethisterone (1000 µg)				5	
Norethisterone acetate (1000 µg)			7		
Lynoestrenol (1000 µg)			3		
Lynoestrenol (2500 µg)					2

[a]   One subject used 500 µg racemic norgestrel (dl-NG).

Table 7
EFFECT OF ORAL CONTRACEPTIVES ON GLUTATHIONE REDUCTASE ACTIVITY

	Untreated controls (A.M. ± S.E.)	Oral contraceptive users (A.M. ± S.E.)
EGR activity (basal) mmol/ℓ hr	107.4 ± 2.6	98.2 ± 2.8 (z)[a]
EGR activity (stimulated) mmol/ℓ hr	125.6 ± 2.7	117.6 ± 2.9 (z)[b]
Activity coefficient	1.18 ± 0.01	1.22 ± 0.02 (z)[b] (F)[a]
Percentage saturation	85.2 ± 0.8	83.0 ± 1.1 (z)[c] (F)[a]
Number	51	63

Note: A.M. ± S.E.: arithmetic mean and standard error. z = z test; F = F test.

[a]   Significant difference ($p < 0.01$).
[b]   Significant difference ($p < 0.025$).
[c]   Significant difference ($p < 0.05$).

Table 8
VITAMIN B$_2$ STATUS: CLASSIFICATION OF SUBJECTS

| | Activity coefficient (%) | | |
	1.00—1.19	1.20—1.39	≥1.40
Controls	66.7	33.3	0
Oral contraceptive users (total)[a]	54.0	33.3	12.7
Low dose[a]	50.0	32.4	17.7
High dose[b]	58.6	34.5	6.9

[a] χ^2 test different from controls ($p < 0.005$).
[b] χ^2 test different from controls ($p < 0.025$).

Table 9
EFFECT OF ORAL CONTRACEPTIVES ON GLUTATHIONE
REDUCTASE ACTIVITY

		Oral contraceptive users	
	Untreated controls	**Low dose**	**High dose**
EGR activity (basal) mmol/ℓ hr	107.4 ± 2.6	97.1 ± 4.1	99.6 ± 3.8
EGR activity (stimulated) mmol/ℓ hr	125.6 ± 2.7	117.0 ± 4.2	118.3 ± 3.9
Activity coefficient	1.18 ± 0.01	1.23 ± 0.02	1.20 ± 0.02
Percentage saturation	85.2 ± 0.8	82.3 ± 1.6	83.8 ± 1.4
Number	51	34	29

Note: Arithmetic mean ± standard error, one-way analysis of variance, and the z and F tests are all statistical methods used on tabulated numbers.

the subjects described as deficient were taking oral contraceptives. Furthermore, it appeared that despite the similarities in the vitamin B_2 status between members of the two subgroups based on dosage, some changes were encountered with greater frequency in the lower dose (30/150) category.

The conclusion reached concerning the dosage of the Pill was tested in another way by making use of a contingency table in an application of χ^2 (Appendix A). The two factors considered were vitamin status (classified into three categories on the basis of activity coefficient) and the type of oral contraceptive (two categories). The outcome of this analysis was that classification of vitamin B_2 status as deficient, low, or acceptable was independent of the type of oral contraceptive. The earlier conclusion was therefore modified to the view that vitamin B_2 status was affected in women taking oral contraceptives of the low 30/150 type and also the higher doses, but there was insufficient statistical evidence to state that one type was worse than another in this respect.

Another variable factor which should be taken into account in a study of this type is the duration of contraceptive pill treatment. An impression was sought by considering three periods of time: (1) up to 6 months, (2) 6 to 12 months, and (3) longer than 12 months. Table 10 summarizes these results with the dosage differentiated, and the same data was used to prepare a composite table. In the first instance, a multivariate analysis was performed using the three defined ranges of activity coefficient with the intention of determining whether this variable was dependent on the other variable, duration of Pill use. Since it had emerged from the analyses of mean and variance that the sample of subjects using low-dose pills was not statistically different from another sample who used higher-dose pills (Table 9), the entire oral contraceptive group was classified according to duration of use and level of activity coefficient. A contingency table of these frequency counts led to the conclusion that the status of vitamin B_2 in oral contraceptive users (based on activity coefficient) was independent of the length of time the Pill had been taken. The heterogeneity of contraceptive type and the relatively small numbers within the higher dose group precluded a multivariate analysis, therefore the conclusions to be drawn are limited. In the lower dose group, which was homogeneous in terms of dosage and type, a similar conclusion was reached for the independence of activity coefficient and duration of use.

This statistical approach was not applied to the enzymatic activity data since ranges have not been established. These would have been arbitrary, and the problem could be solved with another method of analysis. When the means for each variable in Table 10 were compared with the control mean, there appeared to be some trends concerning the effect of duration of usage of low- and high-dose pills on basal and stimulated activity and activity

Table 10
EFFECT OF ORAL CONTRACEPTIVES ON GLUTATHIONE REDUCTASE ACTIVITY

	Duration of treatment (months)					
	Low dose[a]			High dose[b]		
	<6	6—12	>12	<6	6—12	>12
EGR activity[c] (basal)	109.4 ± 7.7	95.3 ± 6.4	93.1 ± 6.9	91.2 ± 5.9	109.7 ± 7.0	99.0 ± 5.5
EGR activity[c] (stimulated)	138.3 ± 6.8	113.5 ± 6.2	110.8 ± 6.8	111.6 ± 5.0	123.5 ± 5.9	118.8 ± 6.1
Activity coefficient	1.28 ± 0.05	1.22 ± 0.04	1.22 ± 0.05	1.24 ± 0.04	1.14 ± 0.03	1.21 ± 0.03
Percentage saturation	78.4 ± 2.6	83.0 ± 2.3	83.5 ± 3.1	81.1 ± 2.7	88.5 ± 2.5	83.1 ± 1.9
Number	7	17	11	6	6	17

Note: Arithmetic mean ± standard error, one-way analysis of variance, and the chi-square test (activity coefficient only) are all statistical methods used on tabulated numbers.

[a] Low dose: no significant difference ($p > 0.05$).
[b] High dose: no significant difference ($p > 0.05$).
[c] mmol/ℓ hr.

coefficient. However, since the variation for each index within each time period was relatively large compared with the variation between periods, these would be generalizations at best. The "t" or "z" test for comparing independent means is invalid in this case, but a statistical test designed to solve this type of problem is analysis of variance (Appendix A), already described in several forms.

One-way analysis of variance showed that the basal activity, stimulated activity, activity coefficient, and percentage saturation of the three treatment periods did not differ when the entire oral contraceptive group was considered, thereby reinforcing the conclusion reached in the preceding analysis. A similar conclusion was reached for the low dose and high dose groups (Table 10).

The overall differences between the samples of women taking oral contraceptives and those who were not taking them can be visualized by reference to Figure 8. Each index, basal and stimulated activity, activity coefficient, and percentage saturation has been shown in parts A, B, C, and D, respectively. The mean for each of the study groups was drawn to scale, so that its relationship to the control mean could be seen more clearly and the range of values for each class is exhibited as the 95% confidence limits. It has already been shown that a relatively high proportion of the test group had an activity coefficient which has been declared low or deficient ($\geqslant 1.20$, Table 8). Since the mean for the control sample was 1.18, there is the question of whether such guidelines should be strictly adhered to, or whether one should be guided by the levels normally encountered in each community. The latter view was considered in two ways: (1) the frequency of individuals who had values below or above the control mean for each parameter was estimated (expressed as percentages in Figure 8), and (2) the individuals whose levels lay outside the 95% confidence limits of the reference sample were identified.

According to the first view, there was a greater proportion of values which were less than the control mean, basal and stimulated activities, and percentage saturation, and greater than the control mean activity coefficient in the total oral contraceptive sample than in the controls (Figure 8). For example, 67% were below the mean basal activity level (compared with

A

Sample	Means and 95 per cent confidence limits	Mean % below	% above
high dose	58.5 — 99.6 — 140.7	72.4**	27.6
low dose	49.0 — 97.1 — 145.2	61.8	38.2
total users	53.5 — 98.2 — 143.0	66.7*	33.3
controls	70.1 — 107.4 — 144.6	49.0	51.0

B

Sample	Means and 95 per cent confidence limits	Mean % below	% above
high dose	76.4 — 118.3 — 160.1	69.0*	31.0
low dose	67.6 — 117.6 — 166.4	70.6*	29.4
total users	71.9 — 117.6 — 163.3	69.8**	30.2
controls	87.0 — 125.6 — 164.2	49.0	51.0

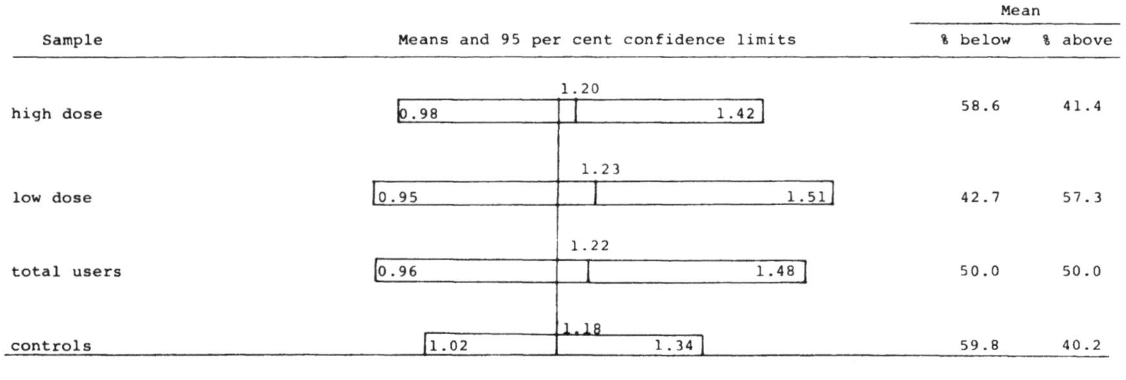

Sample	Means and 95 per cent confidence limits			Mean	
				% below	% above
high dose	0.98	1.20	1.42	58.6	41.4
low dose	0.95	1.23	1.51	42.7	57.3
total users	0.96	1.22	1.48	50.0	50.0
controls	1.02	1.18	1.34	59.8	40.2

C

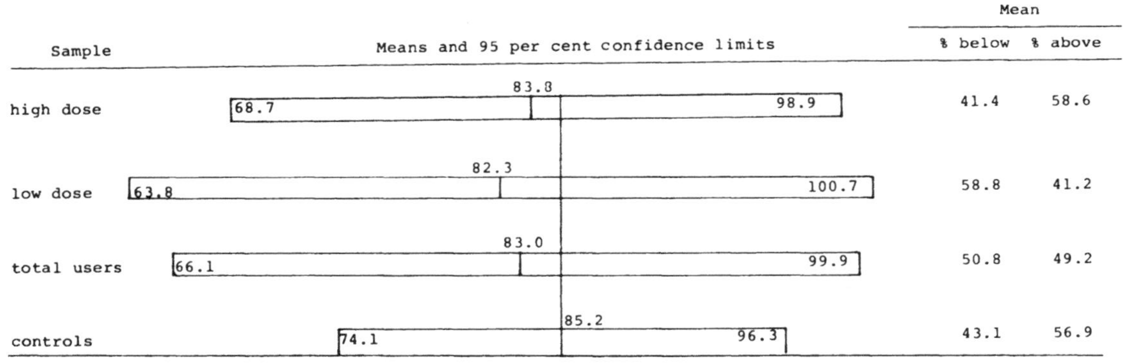

Sample	Means and 95 per cent confidence limits			Mean	
				% below	% above
high dose	68.7	83.8	98.9	41.4	58.6
low dose	63.8	82.3	100.7	58.8	41.2
total users	66.1	83.0	99.9	50.8	49.2
controls	74.1	85.2	96.3	43.1	56.9

D

FIGURE 8. Distribution of results about the control mean (χ^2 test). (A) Glutathione reductase activity (basal - EGR activity, mmol/ℓ hr) (different from controls: $*p < 0.025$. $**p < 0.005$); (B) glutathione reductase activity (stimulated - EGR activity, mmol/ℓ hr) (different from controls: $*p < 0.05$, $** p < 0.005$); (C) activity coefficient (no significant difference: $p > 0.05$); (D) percentage saturation (no significant difference: $p > 0.05$).

49% of the control values), and 50% were above the control mean activity coefficient, compared with 40% of the controls. χ^2 tests showed that the distribution of the control population differed from the oral contraceptive population with respect to basal activity ($p < 0.025$) and stimulated activity ($p < 0.005$), but not activity coefficient or percentage saturation. When the group was subdivided on the basis of dose level, the proportion of values below the mean stimulated activity was substantially greater in both subgroups than the controls ($p < 0.05$), but for basal activity this was the case for the higher dose subgroup only ($p < 0.005$). However, similar consideration of the ratios showed that in this respect both the higher and lower dose groups resembled the controls. Comparisons made between the two types of Pill showed that the distributions of activities and ratios were similar. These observations reinforced the conclusions reached in the preceding analyses indicating that substantial changes in vitamin B_2 status occurred in women taking the oral contraceptive Pill. In terms of the activity coefficient as the only index of vitamin B_2 status, the deleterious effect could not be discerned, whereas the distribution of values of basal and stimulated activity were different from the control profile. These results confirmed the view that total reliance should not be placed on the ratio, activity coefficient.

The proportion of values above and below the average was a means of comparing test and reference samples using information derived from the population in question rather than applying statutory numerical limits, but did not reveal the abnormal results. These were identified in an extension of this approach by ascertaining the values which lay beyond the 95% confidence limits ($\pm$ 2 S.D.) of the control sample. The list displayed in Table 11 was derived from the ranges given in Figure 8. Although values exceeding both upper and lower limits have been included, comments have been restricted to those subjects showing signs of vitamin B_2 depletion, i.e., low basal activity or high activity coefficient. With respect to the latter, only 3.9% of the controls were in this category, whereas there were 18.8% in the entire pill-taking group. Of the latter number, two thirds had been taking the low-dose pill for at least 6 months (with one exception). The remaining one third had all taken a higher dose for 12 months or more. None of the control subjects exhibited low basal activity (defined as being less than the 95% confidence limit), but this existed in 12.5% of the oral contraceptive group. Three fourths of these belonged to the low dose subgroup, with at least 6 months medication. The remaining one fourth, which were in the higher dose group had also been taking their pills for at least 6 months.

Perusal of the contents of Table 11, in particular the observations which have an asterisk, indicates that low basal activity and high activity coefficient are not necessarily found together. This has already been demonstrated by correlation analysis in Section III.B.3, and therefore is a reminder that the identification of a B_2 deficiency by reference to the activity coefficient alone could be misleading. This is exemplified by observation numbers 12, 13, and 24. An explanation of these results can be made by looking at the levels of stimulated activity. Unusual levels of stimulated or total activity were also noted, but since there was no correlation between the magnitude of stimulated activity and activity coefficient (Section III.B.3) this has not been regarded as an indication of vitamin B_2 status. The high values of total activity encountered were all associated with high basal activity, resulting in acceptable ratios. The low levels were usually also combined with low basal activity, but in some cases this produced a normal ratio, as in the women represented by observation numbers 12, 13, and 24. In physiological terms, this might suggest that a state of vitamin B_2 depletion had existed for some time, since under these circumstances there may be loss of apoenzyme GR in response to vitamin B_2 deprivation and concomitant low basal and total activity of the enzyme.

The identification of exceptional observations provided further evidence to support the opinion that the risk of developing vitamin B_2 deficiency is greater when oral contraceptives are being taken, and seemed more likely to occur with a Pill containing 30 μg ethynylestradiol

Table 11
SUBJECTS WITH VALUES OUTSIDE THE 95% CONFIDENCE LIMITS OF CONTROLS

	EGR activity[a]				Comments		
No.	Basal	Stimulated	Activity coefficient	Percentage saturation	Type (μg)		Duration (weeks)
1	*145.3*	163.8	1.13	88.5		Control	
2	71.5	*82.4*	1.15	87.0		Control	
3	71.4	*83.9*	1.18	84.8		Control	
4	142.3	*167.8*	1.18	84.8		Control	
5	104.7	99.3	*0.94*	*106.4*		Control	
6	92.1	125.5	*1.36*[b]	*73.5*		Control	
7	82.5	112.3	*1.36*[b]	*73.5*		Control	
8	66.2[b]	93.1	*1.41*[b]	*70.9*	30 EE, 150 NG		9
9	57.8[b]	76.8	*1.34*[b]	*74.6*	30 EE, 150 NG		10
10	40.7[b]	*61.9*	*1.54*[b]	*64.9*	30 EE, 150 NG		12
11	60.3[b]	86.0	*1.43*[b]	*69.9*	30 EE, 150 NG		30
12	64.6[b]	76.6	1.19	84.0	30 EE, 150 NG		18
13	68.6[b]	*83.9*	1.23	81.3	30 EE, 150 NG		18
14	142.7	*166.2*	1.17	85.5	30 EE, 150 NG		12
15	78.7	115.4	*1.48*[b]	*67.6*	30 EE, 150 NG		5
16	80.3	109.6	*1.36*[b]	*73.5*	30 EE, 150 NG		11
17	70.9	101.0	*1.42*[b]	*70.4*	30 EE, 150 NG		6
18	116.3	119.4	1.03	*97.1*	30 EE, 150 NG		6
19	109.4	109.5	*1.00*	*100.0*	30 EE, 150 NG		18
20	94.6	141.8	*1.51*[b]	*66.2*	30 EE, 150 NG		36
21	123.4	125.5	*1.02*	*98.0*	30 EE, 150 NG		36
22	143.7	137.3	*0.96*	*104.2*	30 EE, 150 NG		60
23	66.9[b]	87.5	1.31	76.3	50 ME, 1000 NET		3
24	66.2[b]	*76.6*	1.16	86.2	50 EE, 1000 NET		39
25	*152.0*	*165.0*	1.09	91.7	50 EE, 1000 NETA		84
26	104.0	144.2	*1.39*[b]	*71.9*	50 EE, 250 NG		24
27	90.0	127.6	*1.42*[b]	*70.4*	50 EE, 250 NG		15
28	103.3	145.1	*1.41*[b]	*70.9*	50 ME, 1000 NET		18
29	75.9	102.3	*1.35*[b]	*74.1*	75 ME, 2500 LYN		18

Note: Abbreviations: EE, ethynylestradiol; ME, mestranol; NG, levo-norgestrel; NET, norethisterone; NETA, norethisterone acetate; LYN, lynoestrenol.

[a] mmol/ℓ hr. Italicized numbers exceed 95% confidence limits.
[b] Basal activity less than or activity coefficient greater than 95% confidence limits.

and 150 μg d-NG than with other combinations. Since equally sized groups could not be obtained for each of the other types of Pill represented, with the additional variable factor of duration of use caution was exercised in expressing the view that particular combinations were responsible for the altered vitamin B_2 status. However, it can be stated that the low-dose Pill is associated with these changes and that they are likely to occur in a significant number of women.

The relationship between basal activity of GR and the activity coefficient was discussed in Section III.B.3, but the absence of deficient subjects in that sample was a disadvantage. In the sample drawn from the population taking oral contraceptives, these occurred at a higher frequency and therefore this aspect could be enlarged upon. Table 12 shows this entire sample, and also two dosage categories within the sample, in terms of basal activity (mean and standard error). These were classified on the basis of the ranges of activity coefficient which already have been defined. A comparison of the means for each range,

Table 12
RELATIONSHIP BETWEEN GLUTATHIONE REDUCTASE ACTIVITY AND ACTIVITY COEFFICIENT IN ORAL CONTRACEPTIVE USERS

	EGR activity (basal)[a]		
Activity coefficient	Low dose[b] (A.M. ± S.E.)	High dose[c] (A.M. ± S.E.)	Total[d] (A.M. ± S.E.)
1.00—1.19	108.7 ± 4.3 (17)	106.7 ± 5.8 (16)	107.7 ± 3.5 (33)
1.20—1.39	94.7 ± 6.9 (11)	88.8 ± 4.2 (10)	91.9 ± 4.1 (21)
≥1.40	68.6 ± 7.4 (6)	96.7 ± 6.7 (2)	75.6 ± 7.2 (8)

Note: A.M. ± S.E.: arithmetic mean ± standard error.

[a]   mmol/ℓ hr.
[b]   One-way analysis of variance ($p < 0.0006$).
[c]   One-way analysis of variance ($p < 0.1006$).
[d]   One-way analysis of variance ($p < 0.0002$).

Table 13
SUMMARY OF RELATIONSHIPS BETWEEN VARIABLES IN THE GLUTATHIONE REDUCTASE TEST

Controls	Oral contraceptive users		
	Total	Low dose	High dose
SA = 0.96 BA + 22.35	SA = 0.92 BA + 27.18	SA = 0.93 BA + 26.41	SA = 0.90 BA + 28.36
AC = −0.002 BA + 1.392	AC = −0.003 BA + 1.545	AC = −0.004 BA + 1.592	AC = −0.003 BA + 1.458
AC = 20.09 BA⁻¹ + 0.99	AC = 25.17 BA⁻¹ + 0.95	AC = 25.59 BA⁻¹ + 0.95	AC = 22.81 BA⁻¹ + 0.97
PS = 0.14 BA + 70.07	PS = 0.22 BA + 61.80	PS = 0.23 BA + 59.54	PS = 0.18 BA + 65.57

Note: Abbreviations: BA, EGR activity (basal); SA, EGR activity (stimulated); AC, activity coefficient; PS, percentage saturation.

using a test of one-way analysis of variance, led to the conclusion that at least two of these were different in the sample considered as a whole and in the low-dose subgroup. The other subgroup was incomplete, with only two individuals in one of the categories. Table 12 also shows that basal activity was greatest in the lowest range of activity coefficient, and least in the highest range. This trend was examined in more detail, with reference to the analogy already drawn between making repeated measurements during progressive depletion of a person and single measurements on a group of people representing a spectrum of nutrition states. Furthermore, as oral contraceptives do cause lowered vitamin B_2 status, this then may be reflected in the regression analysis. In theory, a decrease in basal activity but not stimulated activity, would result in a displaced regression line having the same gradient, but a larger intercept.

The correlation/regression analysis describing the data obtained from the reference sample (Section III.B.3) was also applied to the test sample. The close relationship between the two activities (basal and stimulated) can be described as linear ($r = 0.89$ to 0.91; $p \sim 0$). Since the confidence limits of the regression lines are wider in these groups, there was no difference according to a statistical comparison. Appendix A describes how this was done. The four regression lines which fitted the data of the oral contraceptive groups and also the control data can be compared by referring to Table 13 (see line 1).

The likelihood of encountering high activity coefficients in association with low levels of basal activity can be assessed visually from the graphs of activity vs. coefficient. These are Figure 9, showing the entire oral contraceptive sample with the control sample for convenience, and Figure 10, in which the individual subgroups can be discerned. The correlation between the two variables in the four groups is summarized in Table 14 (see column 3).

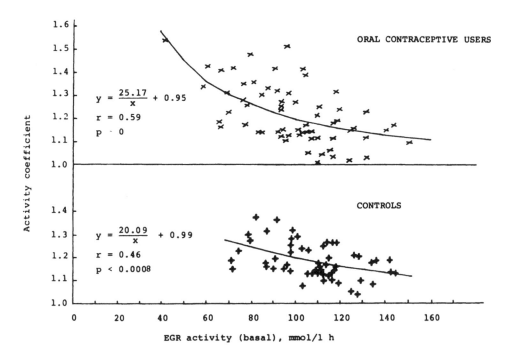

FIGURE 9. Relationship between glutathione reductase activity (basal) and activity coefficient.

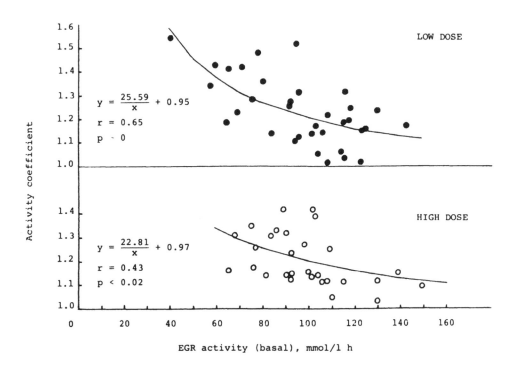

FIGURE 10. Relationship between glutathione reductase activity (basal) and activity coefficient in a sample of oral contraceptive users.

Table 14
SUMMARY OF CORRELATIONS BETWEEN VARIABLES IN THE GLUTATHIONE REDUCTASE TEST

	EGR activity (basal)	EGR activity (stimulated)		Activity coefficient		Percentage saturation	
		r	r²	r	r²	r	r²
Controls (n = 51)	BA	0.93[a]	0.86	−0.47[b]	0.22	0.47[b]	0.22
	1/BA			0.46[c]	0.21		
Oral contraceptive users (total) (n = 63)	BA	0.90[a]	0.81	−0.57[a]	0.33	0.57[a]	0.33
	1/BA			0.59[a]	0.35		
Low dose (n = 34)	BA	0.91[a]	0.83	−0.63[d]	0.40	0.61[d]	0.37
	1/BA			0.65[a]	0.42		
High dose (n = 29)	BA	0.89[a]	0.79	−0.47[e]	0.22	0.50[f]	0.25
	1/BA			0.43[g]	0.18		

[a] $p \sim 0$.
[b] $p < 0.0005$.
[c] $p < 0.0008$.
[d] $p < 0.0001$.
[e] $p < 0.01$.
[f] $p < 0.006$.
[g] $p < 0.02$.

Two functions were included in the table, notwithstanding that the reciprocal was more logical (Section III.B.3), since in the high-dose group the r values differed and therefore did not conform. It had been shown with the control data that despite a highly significant statistical correlation, a limitation existed with respect to the practical application since r² was only 0.22. That is, a given value of activity coefficient was not a very reliable means of finding the level of enzymatic activity. However, in the low dose (30/150) subgroup, the correlation was higher. For the latter r was 0.65, and therefore r² was 0.42, almost double that previously obtained. However, for predictive purposes this could not be regarded as very high. The four regression curves summarized in Table 13 (see lines 2 and 3) were similar as there was a wider scatter in the test groups. These graphs illustrate the point that the highest ratio is not necessarily associated with low basal activity and that a normal activity coefficient can be encountered at the lower end of the scale.

A graph of percentage saturation with cofactor against basal activity of GR gave rise to similar observations. Tables 13 and 14 describe the positive linear correlation to be found between these two variables. In the control sample the minimum percentage saturation was 74, but it was expected that this would be lower in the test sample. Sixty-five percent (65%) saturation was attributed to one member of the low-dose group. In view of the possible decrease in total or stimulated activity, it is not known what the lower limit would be. During enforced depletion, this can fall to about 50% but it may not be valid to extrapolate this result to the general population since changes, if they occur, are likely to take place over a longer period of time.

IV. DISCUSSION

The erythrocyte glutathione reductase test, introduced about 12 years ago[26,31,37] emerged from an interest in the biochemical assessment of sub-clinical riboflavin deficiency. In principle, the method used now is essentially similar. However, there have been many other published accounts of the characteristics of the reaction,[32,42,43,45,46] with differing recommendations for achieving the optimum conditions in the system. An examination of one of those which were originally described[32,43] was commissioned by the World Health Organization (Project on Standardization and Quality Control of Laboratory Methods for the Assessment of Vitamin Status). This was prompted by a desire to determine the best means by which B_2 nutriture could be assessed. Since this test was favored for many circumstances, it was logical to decide on one particular experimental version. Furthermore, this would facilitate interlaboratory comparisons, which are likely to yield closer results when standardized methods are used.[47] The revised version was applied to the present study.

During the revision stage, the preparation of the erythrocyte enzyme was improved. The cells had been lysed shortly after washing the blood. If the assay could not be performed immediately, then the lysate was frozen. Retention of the erythrocytes intact for as long as possible, which entailed freezing them immediately after washing and only inducing lysis when required, was believed to be one practice which prevented an appreciable change in activity. Others[45,48] have found that GR is reasonably stable under various conditions, but should be measured within about 30 days. It was not usually necessary to store specimens for longer than 1 week in the present study. After about 4 weeks, it was noticed that an insoluble material appeared in the hemolysate. This has been reported[45] not to affect the FAD stimulated activity, but may decrease the absolute enzyme activity.

During the course of an evaluation of GR activity,[49] it was found that prior dilution of the erythrocyte enzyme caused a decrease in activity to occur during an incubation of the reaction mixture for 2 hr at 37°C, but this did not occur if the sample was not prediluted. However, in serum, inactivation was retarded by diluting it. A number of the studies recorded in the literature have been based on measurements of washed blood which was lysed on dilution and frozen prior to analysis, making it difficult to make comparisons.

One study of the assay[37] showed that several other factors affected the results. Within 4 hr of hemolysis, the FAD effect remained fairly constant, but activity showed signs of change. It was also found that the volume of the hemolysate and the duration of storage could alter the activity, but the activity coefficient was less susceptible. It was therefore recommended that the absolute values of enzymatic activity should be interpreted with care. Efforts were made to control these variables. It was found that agents which resulted in changes to both basal and stimulated activities had less effect on the activity coefficient. However, with respect to random error, the outcome varied. If a biased error occurred and it affected both the basal and stimulated activities, the consequences could be masked in the activity coefficient. The errors may be magnified when it is envisaged that the ratio y/x could be obtained from measurements wherein the error was $(y + \Delta y)/(x - \Delta x)$. For example, if y = 120 and x = 100, with errors of 5%, the error in the activity coefficient could be approximately 10%.

The order in which the reactants are added seems to have an effect on the final result. In some methods, glutathione initiated the reaction,[45,46] but these activities were lower than when other methods were used. It was more satisfactory to start the reaction with NADPH.[28,42,50]

It has been demonstrated that the hydrolysis of FAD which was added to whole blood could be attributed mainly to plasma, and that of FMN to red cells.[51] Therefore, the in vitro concentration of FAD is not markedly affected by the incubation with red cell lysates. One hour was believed to be necessary to allow union with the apoenzyme, but a lesser time sufficed.[42] The combination of GR and FAD is irreversible, according to the results of attempts to remove the cofactor by dialysis or dilution.[32]

The problems associated with enzyme determinations were under discussion recently,[47] and one advantage of this assay is the generation of dimensionless data, in this case, the expression activity coefficient. This also avoided the question of the accuracy of the technique, which was difficult to judge because the results could not be compared with the absolute level of the substance measured.

Judgment of the results relied rather heavily on the application of statistics. One of the impressions which was gained from the contemplation of other similar studies was that this aspect was not well documented and that there appeared to be heavy reliance on the student *t* test to differentiate between small and selected groups. Implicit in the use of this statistic is the assumption that levels of the sample variable are distributed with normal frequency in the parent population. However, there have been at least two surveys of nutritional factors wherein it was demonstrated that the frequency distribution of many of these variables was not normal.[52,53] It was claimed that the concentrations of many circulating vitamins (B_{12}, B_6, B_1, B_2, nicotinic acid, biotin, pantothenic acid, and folic acid) conformed to a logarithmic-normal pattern[52] as did folic acid in cerebrospinal fluid.[53] It was pointed out that if this was disregarded, then erroneous interpretations of the data would ensue. If a logarithmic-normal distribution could be demonstrated, then statistical analysis by methods such as the *t* test should be based on the logarithm of the primary value.[52,53] When the distribution is not known, then nonparametric statistics can be used.

The concentration of riboflavin in serum exhibited a logarithmic-normal frequency distribution.[52] It is generally assumed that the erythrocyte basal GR activity reflects the serum level. However, this criterion of vitamin B_2 status has not been defined in a statistical sense. There was evidence for a lognormal distribution of basal activities (Section III.B.1) but it was not conclusive. Either a normal or skewed pattern appeared to be applicable in another study[54] judging by the displayed histogram. Although a lognormal distribution is skewed, the converse is not axiomatic. For the stimulated activity it was found that a normal distribution could be used. Some histograms of activity coefficients[37,54,55] looked fairly symmetrical whereas another was skewed.[34] One group[55] concluded that the frequency distribution was markedly skewed and a lognormal rather than a Gaussian distribution best fitted the data. Others[37] believed that there was not a significant departure from normality using a χ^2 test of goodness of fit.

The value of enzymatic activity as a measure of vitamin B_2 status can be justified, but its limitations were apparent when an attempt was made to judge the results in terms of those from other laboratories. More often than not, these values were not given and a lack of uniformity concerning units and actual assay procedure also precluded a comparison on that basis. In this context, the unifying approach of the World Health Organization toward the GR methods was welcomed. Some figures have emerged from this work, but most of these are unpublished at this stage, and furthermore refer to countries which are quite different from Australia in terms of race and diet, making it difficult to compare the results. Some average basal activities (mmol/ℓ hr) were 66 and 77 (Bombay) and 70 (Hyderabad), which are in the low region by Australian standards. The activity coefficients were 1.56, 1.43, and 1.78, respectively (unpublished WHO nutrition—oral contraceptive interaction studies). However, higher levels were found in a reference population and these were similar to those found in the present study. In a published report of a study conducted in Bombay[56] the reference females had mean activities of 97.6 (basal) and 116.0 (stimulated) which were a little lower than those recorded for the Australian study (107.4 and 125.6). The merit of the activity coefficient is illustrated when it is realized that the ratio of activities was similar (1.19 and 1.18, respectively). It is this expression which is customarily used, but dimensionless data does not obviate the need for standardization in methodology.

The existence of published guidelines for reference ranges of activity coefficient defining vitamin B_2 status implies that it should not be necessary for each laboratory to establish their own, and those exhibited in Figure 4 have been relied upon in many studies for judging acceptable, low, and deficient states without reference to the local averages. These ranges emanated from controlled human riboflavin deficiency studies (Section I.E.). The allowance of 0 to 20% stimulation of activity with FAD (activity coefficient 1.00 to 1.20) was believed to provide a means of identifying the marginally deficient subject and gave a normal range. It was apparent that even in the latter, the enzyme was not completely saturated with FAD, and would respond to supplements of riboflavin. Another laboratory differed, defining adequate as 1.00 to 1.20, marginal as 1.20 to 1.25, and deficient as greater than 1.25.[54] Limits of greater than 1.29,[50] greater than 1.30,[45] greater than 1.25,[26] or even greater than 1.40[23] have also been quoted as indicative of B_2 deficiency. Since there are several opinions as to what constitutes a normal activity coefficient, it would seem that until the methodology becomes more uniform, each laboratory should be guided by ranges established from its own results.

The results were viewed in relation to the ranges of activity coefficient given in Figure 4, as well as by limits determined from the sample. The latter, being those which exceeded the mean plus 2 S.D., were a minority of the sample. According to the defined ranges, none were deficient (≥ 1.40), but 33% had low or marginal status (≥ 1.20). In one of the first applications of the enzymatic technique (to an American high school population), 10.8% of caucasian girls exceeded 1.20,[34] whereas in a British study there were none.[57] In comparison with some of the averages of reference populations in other Western countries,[50,54] the results obtained were reasonable. However, the activity coefficient was not entirely independent of the method. When lower ratios were recorded[34,37,57] these were associated with ranges which were physiologically illogical, such as 0.80 to 1.13. These resulted in a mean of 1.01,[57] or the normal range proposed[37] which gave an upper limit of 1.20, but extended down to 0.90. These would not be suitable for comparisons. This appears to be a function of the analytical method, attributed to an excessive concentration of FAD, which causes inhibition of enzyme activity.[55] It was not evident in this study, since the results ranged from 1.02 to 1.34. Higher mean activity coefficients of 1.38 (Australians),[58] 1.32 (British),[42] and 1.33 (Koreans)[48] have also been quoted and upper limits of normal were given as 1.76[42] and 1.61.[48]

An adequate diet is not necessarily that required for maintaining the enzyme at maximum saturation with cofactor. This is exemplified by the finding that 73 to 98 percentage saturation of the enzyme with FAD satisfied the physiological requirements for health. Oral vitamin supplementation to healthy subjects on unrestricted diets resulted in an increase in saturation for the enzyme, but these higher values (or alternatively, decreased activity coefficients), should not be regarded as reference values.

A high correlation between basal activity and stimulated activity was noticed during one of the first descriptions of the characteristics of the GR test.[37] A group of selected subjects who had activity coefficients of more than 1.20 conformed to a regression line which could be extrapolated to a y intercept (stimulated activity), but after administration of riboflavin, the line passed through the origin with a similar gradient of approximately 1. That is, the ideal vitamin status (activity coefficient of 1.0 or 100% saturation) could be achieved. It could be envisaged that in each person there was an increase in the amount of FAD bound to the enzyme (reflected in the increase in basal activities) until the maximum was reached; the limit was set by the total amount of enzyme. The regression line was therefore displaced to the right. Closer examination of the data indicated that in some individuals there must have also been synthesis of apoenzyme, since the range of stimulated activity changed from 0.100 to 0.250Δ to 0.220 to 0.320 ΔA. The ratio of 1 was also commonly encountered in the reference sample of that study, but should not necessarily be interpreted as reflecting complete saturation in vivo, since the mean represented a range of 0.84 to 1.24. This can be attributed to suppression of stimulated activity. Viewed in the context of subsequent work, it is more likely that GR is not usually fully saturated even in subjects taking unrestricted, unsupplemented diets. Therefore, the regression line displayed for basal and stimulated activities in the control sample was considered to be reasonable.

The presence of a y (stimulated activity) intercept after extrapolation meant that the ratio of stimulated activity to basal activity was related to the magnitude of the basal activity. This was also demonstrated by plotting all of the basal activities in the sample against the respective activity coefficients. In general, as basal activity tended toward zero, activity coefficient was higher, whereas it approached a minimum at infinite activity. Others[31,50] also found that to a high activity coefficient corresponded a low basal activity, and those designated deficient had significantly lower enzyme activities than those who were not deficient. This relationship has hitherto not been explored. It has been assumed, probably validly, in individuals who were deprived of riboflavin in a trial and monitored by the increasing activity coefficient over a period of time. This response has been applied to other situations. However, the same relationship was not assumed in the study undertaken of the type in which the ranges of basal activities and activity coefficients were obtained from many different people in natural circumstances rather than from an individual responding to the exclusion of one vitamin and measured repeatedly.

There were examples in the reference sample where a low activity coefficient occurred at the lower end of the activity scale, casting doubt on the reliability of this ratio as the sole indicator of riboflavin nutriture. Vitamin B_2 status is usually quoted in terms of the activity coefficient and the judgment made on this basis without regard to the contributing enzymatic activities, since it is usually assumed that if the ratio is above a particular value then low basal activity is responsible; the latter was initially caused by a decrease in the concentration of vitamin B_2. However, circumstances have been recognized which can lead to concomitant lowering of the total enzyme concentration. Thus, it is conceivable that the activity coefficient could be within the normal range, and hence not extraordinary. The results showed that a very significant statistical correlation (r) may not have such a good practical application (r^2). Caution was therefore exercised in evaluating the results and total reliance was not placed on the activity coefficient. Notwithstanding this problem, the basal activity *per se* should be monitored since it measures the amount of FAD bound to GR and is also a reflection of the amount of FAD in the cells, and should not fall too low. Even if the percentage of the enzyme saturated is fairly high, the total amount of FAD may be insufficient.

An upper limit for the activity coefficient is difficult to define. During depletion experiments this has reached 2.34[28] which is equivalent to 43% saturation. These types of investigations are, of necessity, short term, and loss of FAD is exhibited in lowered basal activity and concomitant high activity coefficient. This apparently has no effect on the enzyme protein (measured as stimulated activity) as exemplified by the rapid return to normal after repletion of the vitamin was instituted. It is, however, relevant to ask whether these results can be extended to situations in the field where other conditions are likely to complicate vitamin B_2 deficiency. It is possible that with the coexistence of additional nutritional shortages and longer term states the apoenzyme may also decrease in concentration, and the activity coefficient could be lower than expected. Remarkably low saturation of about 38% (activity coefficient of 2.64) has, however, been recorded in a study of Thai villagers[59] and also during an extensive survey carried out by the World Health Organization in India and Thailand.

Some of the results obtained by others when vitamin B_2 status was assessed in women who were using oral contraceptives have already been discussed.[1] A deterioration was recorded in many of these, i.e., decreased enzymatic activity (basal) and increased activity coefficient, but in some studies there was no significant effect.[60-63] In the sample of young Australians, a cross-sectional approach demonstrated a difference in vitamin B_2 status from various points of view, using the GR test, and it did appear to be adverse. The activity coefficient increased by a fairly small amount, from a reference value of 1.18 to 1.22, and this was significant. One reason why some others have failed to discriminate between samples in cross-sectional studies could be a statistical one, since large numbers are required to demonstrate small differences. This may apply to the evaluations of Israeli or American women.[60-62]

In general, it appears that a deterioration in vitamin B_2 status can be consistently demonstrated by the GR test in oral contraceptive users living under poor socioeconomic conditions. Low income was reflected in a poorer diet and in a lower concentration of riboflavin in the serum.[64] Financial status was believed to be a factor responsible for generally higher activity coefficients in samples of women from India and Thailand (unpublished WHO nutrition-oral contraceptive interaction studies). The contributing role of this factor has been an argument used to explain why riboflavin nutrition in women of better socioeconomic status did not appear to be affected by oral contraceptive usage. The implication is that significant depletion of riboflavin body stores occurred only when marginal dietary intake was coupled with Pill use.[62] It could be argued, however, that oral contraceptive agents are capable of eliciting similar depletion in all of the subjects, irrespective of socioeconomic factors, but that it is detected more easily in those who are at the lower end of the B_2 spectrum initially. This is based on a consideration of the relationship which was found between basal and stimulated activity in the reference sample. Disregarding the biological meaning for the moment, in mathematical terms it was linear ($y = ax + b$) with a positive y intercept, and it can be proved that an incremental change in x (Δx) has a greater effect on $\Delta(y/x)$ at the lower extremity of the x axis than in the upper regions. Physiologically, this means that a change in basal activity would result in a greater alteration to the activity coefficient if it took place when basal activity was low (deficient), than if this change occurred in a person who had a high (normal) basal activity. In fact, for the same change of activity coefficient to be detected in those with low and high basal activity, there would need to be a much greater alteration in basal activity in those who had high activity initially. In this case the activity coefficient is a less sensitive criterion of B_2 status than basal activity, and unless large numbers are involved a difference may be insignificant. Generally, basal activities were not quoted, but may be valuable on the basis of this reasoning.

The failure to find a significant change in the GR test in severely malnourished women on the Pill[65,66] may be related to the recognition of a minimum concentration of riboflavin which can be reached, this being facilitated by the conservation of FAD for its cofactor role when a supply of the vitamin is limited (Section I.C.).

If the effect of oral contraceptives is to cause a depletion of riboflavin in the classical sense, then it should be reflected in the graphical analysis which was attempted. A decrease in basal activity, but not the total or stimulated activity, would affect the regression line connecting these two by increasing the y intercept but not the gradient, that is, displacement to the left. The question has not been answered elsewhere, but is analogous to the outcome of vitamin repletion of deficient subjects.[37] This would be more easily tested in the follow-up approach. In the cross-sectional study the normal scatter and heterogeneity of dose and treatment time made it difficult to test this hypothesis. The gradients of all of the samples were similar, and the intercepts were greater for the study samples than for the reference, but these differences were not statistically significant.

The conclusions concerning the high dosage group were conservative since it was not homogeneous. It was, however, apparent from the results that the low dose Pill was implicated in a deterioration of vitamin B_2 status. No significant difference was noted by others[57] in women taking the low dose 30/150 Pill after 3 months (follow-up study) or after longer medication. Where an adverse effect was recorded, oral pills containing 30 or 50 μg ethynylestradiol and 150 μg d-NG did not differ significantly in the type or magnitude of the change (unpublished data, World Health Organization).

On the basis of results derived from animal experiments, oral contraceptive steroids probably do not elicit a deficiency of the type traditionally encountered during dietary restriction, but are responsible for a relative state of depletion caused by a redistribution of riboflavin cofactors.[67] This could be due to extra demands by some tissues, in particular, the liver. The explanation was empirical, and whether the results in rats can be extrapolated to humans remains speculative. However, the question of why oral contraceptives should affect riboflavin status can be addressed from another, theoretical point of view. This is

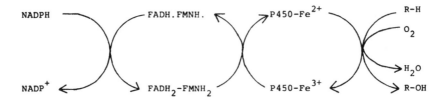

FIGURE 11.   Scheme showing role of flavoprotein in liver microsomal hydroxylation reactions.

based on the premise that a regular dose of the synthetic steroids in the Pill must be catabolized, and among the enzymes which carry this out in liver are hydroxylases. They catalyze the insertion of an hydroxyl group using molecular oxygen as a source and involve electron transfer from NADPH to cytochrome P_{450}. Common to these systems are flavoproteins, containing FMN and FAD.[68-70] Figure 11 depicts a scheme by which hydroxylation is believed to occur. It could be argued that there would be enhanced hepatic requirements for riboflavin under these circumstances, to accommodate the additional enzyme which is needed to metabolize the steroids in oral contraceptives.

The metabolism of each of the commonly used steroids was considered from published accounts, and it was apparent that initial hydroxylation, catalyzed by the above-mentioned complex, is significant. The pharmacokinetics have been reviewed recently[71-73] and the salient features have been summarized.[74] The metabolism of the two estrogens, namely ethynylestradiol and mestranol, appears to be fairly well understood. Mestranol is converted to ethynylestradiol and, apart from de-ethynylation, the main metabolic pathways involve hydroxylation to give 2-hydroxy- and 16β-hydroxy-ethynylestradiol. The estrogen component of combined oral contraceptives could therefore elicit an increase in the amount of flavoprotein complex required and this might be dose related. Dosage did not appear to be a significant factor in determining the disturbance to riboflavin status, and therefore attention was focused on the nature of the progestogen. All of the low-dose estrogen pills contained d-NG. Among the metabolic pathways identified which proceed by way of the cytochrome P_{450} system were 16β-hydroxylation and 2α-hydroxylation.[72,73] The other 17α-ethynyl progestins which are metabolized as norethisterone, and which were only marketed with a high dose (>30 μg) of estrogen, do not appear to undergo this type of hydroxylation, being limited to reductive reactions of ring A.

It was deduced that the potential advantage to be gained by reducing the dose level of estrogen might be counteracted by combining it with norgestrel. In terms of the metabolism, the higher dose of estrogen need not be a disadvantage with norethisterone or a relative of this progestogen. The net effect of various combinations on the hepatic enzymes and ultimately riboflavin status would be difficult to predict accurately, partly because there is marked individual variation in the rate of metabolism. It is worth noting that the l-enantiomer of norgestrel, which is biologically inactive, is metabolized differently from the d-enantiomer and mainly involves hydroxylation at C16.[72]

APPENDIX A: STATISTICS

It was pointed out in the text when statistical methods had been used and, apart from general descriptions, the type and manner with which these were executed was excluded and contained in this appendix. All of these statistics have therefore been compiled, and their method of calculation can mainly be found in the references following. For some of the topics, it was felt that elaboration would be helpful. The calculation of several of the statistics was aided by a computer program called Statpack, compiled by Western Michigan University, Kalamazoo.

1. Error analysis
   Precision, calculated from the differences between paired observations[75,76]
   Coefficient of variation[76]
2. Measures of central tendency
   Arithmetic mean[77]
   Geometric mean[78]
3. Measures of dispersion
   Arithmetic standard deviation and error[77]
   Geometric standard deviation and error[78]
4. Frequency distributions
   Normal distribution[77]
   Logarithmic-normal distribution[79]
   Testing for non-normality of distribution. Two methods were used to find out whether a distribution could be described as normal: (a) A comparison was made between the observed frequency distribution and that which would be expected for a sample if it was normally distributed, the outcome being quantitated by the χ^2 test. In order to do this, the sample was first grouped into class intervals of the variable and this provided the observed frequency of observations. The variable was converted into a standardized variable, $\hat{z}$, whose distribution has a mean of zero and variance of one, based on the concept of the standardized normal distribution.[77] Tables depicting the probability areas under regions of this curve were used to calculate the expected frequency for each class interval. The observed and expected frequencies were then compared by means of a χ^2 test.[77] (b) The standardized normal distribution was subdivided into areas of equal probability and the expected frequency of observations to be found within each of these areas was compared with the observed frequency found in the corresponding transformed class intervals, using the χ^2 test.
5. Inferences concerning two populations
   Inferences concerning the difference between two independent means (large samples); z statistic[77]
   Inferences concerning two variances; F statistic[77]
   Inferences concerning the difference between two independent means (small samples); t statistic (parametric) and Mann-Whitney U test (nonparametric)[77]
   Inferences concerning two dependent means; paired t test (parametric)[77] and Wilcoxon signed-rank test (nonparametric)[80]
6. Inferences involving enumerative data; χ^2 statistic
   Inferences concerning multinomial experiments[77]
   Inferences concerning contingency tables[77]
7. Analysis of variance; F statistic
   One-way classification[80]
   Two-way classification; randomized block design[80]
   Three-way classification; factorial experiment[80]
8. Multiple comparison procedures; Tukey's W[80]
9. Linear correlation and regression analysis
   Linear correlation analysis; r statistic (parametric) and Spearman rank correlation (nonparametric)[77]
   Linear regression analysis[77,80]
   Inferences concerning the average value of y.[80] The confidence interval (95%) for the mean value of y was calculated for a given setting of the independent variable x within the range of data available, and the endpoints were plotted to give the confidence bands. For the prediction equation $\hat{y} = ax + b$ the confidence interval is given by the formula

$$\hat{y} \pm t_{\alpha/2} \, s \, \sqrt{\ell'(X'X)^{-1}\ell}$$

where X = matrix for all settings for x;

ℓ = matrix of a given setting of X

Comparing the slopes of regression lines[80]

Comparing the intercepts of regression lines. The same principles as above can be applied after it has been determined whether there is any difference in slopes.

REFERENCES

1. **Tonkin, S. Y.**, Oral contraceptives and vitamin status, in *Vitamins in Human Biology and Medicine*, Briggs, M. H., Ed., CRC Press, Boca Raton, Fla., 1981, 29.

2. **Chen, C. and Yamauchi, K.**, Histochemical study on riboflavin, *J. Vitaminol.*, 7, 163, 1961.

3. **Rivlin, R. S.**, Riboflavin metabolism, *N. Engl. J. Med.*, 283, 463, 1970.

4. **McCormick, D. B.**, The fate of riboflavin in the mammal, *Nutr. Rev.*, 30, 75, 1972.

5. **Baker, H., Frank, O., Feingold, S., and Leevy, C. M.**, Vitamin distribution in human plasma proteins, *Nature (London)*, 215, 84, 1967.

6. **Jusko, W. J. and Levy, G.**, Plasma protein binding of riboflavin and riboflavin-5'-phosphate in man, *J. Pharm. Sci.*, 58, 58, 1969.

7. **McCormick, D. B.**, Riboflavin, in *Present Knowledge in Nutrition*, 4th ed., The Nutrition Foundation, New York, 131, 1976.

8. **Hemmerick, P. and Nagelschneider, G.**, Chemistry and molecular biology of flavins and flavoproteins, *FEBS Lett.*, 8, 69, 1970.

9. **Beutler, E.**, The effect of flavin coenzymes on the activity of erythrocyte enzymes, *Experientia*, 25, 804, 1969.

10. **Yubishi, T., Matsuki, T., Tanishama, K., Takeshita, M., and Yoneyama, Y.**, NADPH-flavin reductase in human erythrocytes and the reduction of methaemoglobin through flavin by the enzyme, *Biochem. Biophys. Res. Commun.*, 76, 174, 1977.

11. **Scott, E. M., Duncan, I. W., and Ekstrand, V.**, Purification and properties of glutathione reductase of human erythrocytes, *J. Biol. Chem.*, 238, 3928, 1963.

12. **Massey, V. and Williams, C. H.** On the reaction mechanism of yeast glutathione reductase, *J. Biol. Chem.*, 240, 4470, 1965.

13. **Prosky, L., Burch, H. B., Bijrablaya, D., Lowry, O. H., and Combs, A. M.**, The effects of galactoflavin on riboflavin enzymes and coenzymes, *J. Biol. Chem.*, 239, 2691, 1964.

14. **Anon.**, The interrelationship between riboflavin and pyridoxine, *Nutr. Rev.*, 35, 237, 1977.

15. *Modern Nutrition in Health and Disease*, Goodhart, R. S. and Shils, M. E., Eds., Lea & Febiger, Philadelphia, 1980.

16. Riboflavin, in *The Vitamins*, Sebrell, W. H. and Harris, R. S., Eds., Vol. 5, 2nd ed., Academic Press, New York, 1972.

17. *Riboflavin*, Rivlin, R. S., Ed., Plenum Press, New York, 1975.

18. **Sharada, D. and Bamji, M. S.**, Erythrocyte glutathione reductase activity and riboflavin concentration in experimental deficiency of some water soluble vitamins, *Int. J. Vitam. Nutr. Res.*, 42, 43, 1972.

19. **Krishnaswamy, K.**, Erythrocyte glutamic oxaloacetic transaminase activity in patients with oral lesions, *Int. J. Vitam. Nutr. Res.*, 41, 247, 1971.

20. **Lakshmi, A. V. and Bamji, M. S.**, Regulation of blood pyridoxal phosphate in riboflavin deficiency in man, *Nutr. Metab.*, 20, 228, 1976.

21. **Bamji, M. S., Sharada, D., and Naidu, A. N.**, A comparison of the fluorometric and microbiological assays for estimating riboflavin content of blood and liver, *Int. J. Vitam. Nutr. Res.*, 43, 351, 1973.

22. **Koziol, J.**, Vitamin B$_2$, in *Methods in Enzymology*, Vol. 18B, McCormick, D. B. and Wright, L. D., Eds., The Nutrition Foundation, New York, 1971.

23. **Sauberlich, H. E., Skala, J. H., and Dowdy, R. P.**, Eds., *Laboratory Tests for the Assessment of Nutritional Status*, CRC Press, Cleveland, 1974.

24. **Pearson, W. N.**, Blood and urinary vitamin levels as potential indices of body stores, *Am. J. Clin. Nutr.*, 20, 514, 1967.

25. **Bessey, O. A., Horwitt, M. K., and Love, R. H.,** Dietary deprivation of riboflavin and blood riboflavin levels in man, *J. Nutr.,* 58, 367, 1956.
26. **Bamji, M. S.,** Glutathione reductase activity in red blood cells and riboflavin nutritional status in humans, *Clin. Chim. Acta,* 26, 263, 1969.
27. **Brubacher, G.,** Biochemical studies for assessment of vitamin status in man, *Bibl. Nutr. Dieta,* 20, 31, 1974.
28. **Tillotson, J. A. and Baker, E. M.,** An enzymatic measurement of the riboflavin status in man, *Am. J. Clin. Nutr.,* 25, 425, 1972.
29. **Kaufmann, N. A. and Guggenheim, K.,** The validity of biochemical assessment of thiamin, riboflavin, and folacin nutriture, *Int. J. Vitam. Nutr. Res.,* 47, 40, 1977.
30. **Metzler, D. E.,** *Biochemistry,* Academic Press, New York, 1977.
31. **Glatzle, D., Weber, F., and Wiss, O.,** Enzymatic test for the detection of a riboflavin deficiency, *Experientia,* 24, 1122, 1968.
32. **Beutler, E.,** Effect of flavin compounds on glutathione reductase activity: in vivo and in vitro studies, *J. Clin. Invest.,* 48, 1957, 1969.
33. **Bamji, M. S.,** Laboratory tests for the assessment of vitamin nutritional status, in *Vitamins in Human Biology and Medicine,* Briggs, M. H., Ed., CRC Press, Boca Raton, Fla., 1981.
34. **Sauberlich, H. E., Judd, J. H., Nichoalds, G. E., Broquist, H. P., and Darby, W. J.,** Application of the erythrocyte glutathione reductase assay in evaluating riboflavin nutritional status in a high school population, *Am. J. Clin. Nutr.,* 25, 756, 1972.
35. **Prentice, A. M. and Bates, C. J.,** A biochemical evaluation of the erythrocyte glutathione reductase test for riboflavin status, *Br. J. Nutr.,* 45, 37 and 53, 1981.
36. **Tillotson, J. A. and Sauberlich, H. E.,** Effect of riboflavin depletion and repletion on the erythrocyte glutathione reductase in the rat, *J. Nutr.,* 101, 1459, 1971.
37. **Glatzle, D., Körner, W. F., and Wiss, O.,** Method for the detection of a biochemical riboflavin deficiency, *Int. J. Vitam. Nutr. Res.,* 40, 166, 1970.
38. **Anon.,** Erythrocyte glutathione reductase — a measure of riboflavin nutritional status, *Nutr. Rev.,* 30, 162, 1972.
39. **Axelrod, A. E. and Trakatellis, A. C.,** Relationship of pyridoxine to immunological phenomena, *Vitam. Horm.,* 22, 591, 1964.
40. **Bamji, M. S.,** Letter to the editor: an enzymatic measurement of riboflavin status in man, *Am. J. Clin. Nutr.,* 26, 237, 1973.
41. **Glatzle, D., Vuilleumier, J. P., Weber, F., and Decker, K.,** Glutathione reductase test with whole blood, a convenient procedure for the assessment of riboflavin status in humans, *Experientia,* 30, 665, 1974.
42. **Bayoumi, R. A. and Rosalki, S. B.,** Evaluation of methods of coenzyme activation of erythrocyte enzymes for detection of deficiency of vitamins B_1, B_2 and B_6, *Clin. Chem.,* 22, 327, 1976.
43. **Beutler, E.,** Glutathione reductase: stimulation in normal subjects by riboflavin supplementation, *Science,* 165, 613, 1969.
44. *Clinical Chemistry: Principles and Techniques,* Henry, K. J., Cannon, D. C., and Winkleman, J. W., Eds., Harper & Row, New York, 1974.
45. **Nichoalds, G. E.,** Assessment of status of riboflavin nutriture by assay of erythrocyte glutathione reductase activity, *Clin. Chem.,* 20, 624, 1974.
46. **Williams, D. G.,** Methods for the estimation of three vitamin dependent red cell enzymes, *Clin. Biochem.,* 9, 252, 1976.
47. **Rosalki, S. B.,** Quality control of enzyme determinations, *Ann. Clin. Biochem.,* 17, 74, 1980.
48. **Tchai, B. S.,** Biochemical assessment of vitamin B_1, B_2 and B_6 nutriture by coenzyme activation on erythrocyte enzymes, *K. J. N.,* 10, 24, 1977.
49. **Spooner, R. J., Delides, A., and Goldberg, D. M.,** Anomalous behavior of glutathione reductase on dilution, *Clin. Chem.,* 22, 1005, 1976.
50. **Hoorn, R. K. J., Flikweert, J. P., and Westerink, D.,** Vitamin B_1, B_2 and B_6 deficiencies in geriatric patients, measured by coenzyme stimulation of enzyme activities, *Clin. Chim. Acta,* 61, 151, 1975.
51. **Okumura, M. and Yagi, K.,** Hydrolysis of flavin adenine dinucleotide and flavin mononucleotide by rabbit blood, *J. Nutr. Sci. Vitaminol.,* 26, 231, 1980.
52. **Ziffer, H., Frank, O., Christakis, G., Talkington, L., and Baker, H.,** Data analysis strategy for nutritional survey of 642 New York city school children, *Am. J. Clin. Nutr.,* 20, 858, 1967.
53. **Weckman, N. and Lehtovaar, R.,** Logarithmic-normal distribution of cerebrospinal fluid folate concentrations, *Experientia,* 15, 585, 1969.
54. **Van den Berg, H., Schreurs, W. H. P., and Joosten, G. P. A.,** Evaluation of the vitamin status in pregnancy, *Int. J. Vitam. Nutr. Res.,* 48, 12, 1978.
55. **Garry, P. J. and Owen, G. M.,** An automated flavin adenine dinucleotide-dependent glutathione reductase assay for assessing riboflavin nutriture, *Am. J. Clin. Nutr.,* 29, 663, 1976.

56. **Joshi, U. M.,** The effects of oral contraceptives on carbohydrate, lipid and protein metabolism in subjects with altered nutritional status and in association with lactation, *J. Steroid Biochem.,* 11, 483, 1979.

57. **Vir, S. C. and Love, A. H. G.,** Riboflavin nutriture of oral contraceptive users, *Int. J. Vitam. Nutr. Res.,* 49, 286, 1979.

58. **Briggs, M. and Briggs, M. H.,** Changes in biochemical indices of vitamin nutrition in women using oral contraceptives during treatment with "Surbex 500", *Curr. Med. Res. Opin.,* 2, 626, 1975.

59. **Amatayakul, K., Sivasomboon, B., and Thanangkul, O.,** Vitamin and trace mineral metabolism in medroxyprogesterone acetate users, *Contraception,* 18(3), 253, 1978.

60. **Lewis, C. M. and King, J. C.,** Effect of oral contraceptive agents on thiamin, riboflavin and pantothenic acid status in young women, *Am. J. Clin. Nutr.,* 33, 832, 1980.

61. **Guggenheim, K. and Segal, S.,** Oral contraceptives and riboflavin nutriture, *Int. J. Vitam. Nutr. Res.,* 47(3), 234, 1977.

62. **Carrigan, P. J., Machinist, J., and Kershner, R. P.,** Riboflavin nutritional status and absorption in oral contraceptive users and nonusers, *Am. J. Clin. Nutr.,* 32, 2047, 1979.

63. **Roe, D. A., Bogusz, S., Sheu, J., and McCormick, D. B.,** Factors affecting riboflavin requirements of oral contraceptive users and nonusers, *Am. J. Clin. Nutr.,* 35, 495, 1982.

64. **Murthy, N. K. and Vijaya, S.,** Vitamin nutritional status of women using oral contraceptive pills, *Ind. J. Nutr. Diet.,* 17, 79, 1980.

65. **Bamji, M. S. and Prema, K.,** Effects of oral contraceptives on vitamin nutrition status of malnourished women, in Abstr. 5th Int. Congr. Horm. Steroids, New Delhi, 1978, 62.

66. **Bamji, M. S., Prema, B. A., Lakshmi, R., Ahmed, F., and Jacob, C. M.,** Oral contraceptive use and vitamin nutrition status of malnourished women - effects of continuous intermittent vitamin supplements, *J. Steroid Biochem.,* 11, 487, 1979.

67. **Ahmed, F. and Bamji, M. S.,** Biochemical basis for the "riboflavin defect" associated with the use of oral contraceptives. A study in female rats, *Contraception,* 14(3), 297, 1976.

68. **Chang, C. K. and Dolphin, D.,** Oxidation and activation by heme proteins, in *Bioorganic Chemistry,* Vol. 4, Van Tamelen, E. E., Ed., Academic Press, New York, 1978, 37.

69. **Ullrich, V.,** Cytochrome P450 and biological hydroxylation reactions, in *Topics in Current Chemistry,* Vol. 83, Dewar, M. J. S. et al., Eds., Springer-Verlag, Basel, 1979, 68.

70. **Gunsalus, I. C. and Sligar, S. G.,** Oxygen reduction by the P450 monooxygenase systems, *Adv. Enzymol.,* 47, 1, 1978.

71. **Helton, E. D. and Goldzieher, J. W.,** The pharmacokinetics of ethynyl estrogens: a review, *Contraception,* 15, 255, 1977.

72. **Ranney, R. E.,** Comparative metabolism of 17-ethynyl steroids used in oral contraceptives, *J. Toxicol. Environ. Health,* 3, 139, 1977.

73. **Briggs, M. H.,** Comparative pharmacodynamics and pharmacokinetics of contraceptive steroids in animals and man: a selective review, in *Clinical Pharmacology and Therapeutics,* Turner, P., Ed., Macmillan, New York, 1980, 493.

74. IARC Monogr. Evaluation Carcinogenic Risk of Chemicals to Humans, Sex Hormones, Vol. 21, International Agency for Research on Cancer, Lyon, 1979.

75. **Youden, W. J.,** *Statistical Methods for Chemists,* John Wiley & Sons, New York, 1951.

76. **Reed, A. H. and Henry, R. J.,** Accuracy, precision, quality control and miscellaneous statistics, in *Clinical Chemistry,* Henry, R. J., Cannon, D. C., and Winkelman, J. E., Eds., Harper & Row, New York, 1976, 287.

77. **Johnson, R.,** *Elementary Statistics,* Duxbury, North Scituate, Mass., 1980.

78. **Patteeuw, J. L.,** Review of mathematics, in The Industrial Environment-Its Evaluation and Control, U.S. Department of Health, Education and Welfare, Washington, D.C., 11, 1973.

79. *Scientific Tables,* Diem, K. and Lentner, C., Eds., Ciba-Geigy, Basel, 1970, 145.

80. **Ott, L.,** *An Introduction to Statistical Methods and Data Analysis,* Duxbury, North Scituate, Mass., 1977.

Chapter 7

CAUSES OF NEURAL TUBE MALFORMATION AND THEIR PREVENTION BY DIETARY IMPROVEMENT AND PRECONCEPTIONAL SUPPLEMENTATION WITH FOLIC ACID AND MULTIVITAMINS

K. M. Laurence

TABLE OF CONTENTS

I. INTRODUCTION

Neural tube defects (NTD) include a wide spectrum of abnormalities all due to the failure of the neural tube to close properly toward the end of the fourth week of development. They are still one of the most common serious malformations and a major cause of stillbirth infant death and childhood handicap, even though their prevalence seems to have dropped in a number of parts of the world.

II. NEURAL TUBE DEFECT SPECTRUM OF ABNORMALITIES

Closure defects at the top end of the neural tube result in anencephaly, a lethal condition generally leading to stillbirth, often prematurely. There is absence of the major part of the vault, and disordered, congested, largely degenerated brain tissue lies exposed on the base (Figure 1). Many cases have an associated spinal rachischisis but some have a separate discreet spina bifida.

Closure abnormality a little further down the neuraxis may result in encephalocele, a relatively uncommon lesion where a defect in the posterior part of the skull or the upper cervical spine allows meninges, often with very disordered brain tissue, to protrude. It is almost always covered totally by skin or thick membrane (Figure 2). Such cases have a fair chance of survival, sometimes with little actual paralysis but the infants generally grow up spastic and often blind and severely retarded.[1]

Failure of complete closure down the remainder of the neuraxis will lead to the common myelocele (myelomeningocele), meningocele, or to complicated spina bifida occulta. The myelocele is a very variable abnormality,[1,2] where most commonly at birth a portion of the unclosed spinal cord lies exposed on the surface as a neural plaque giving the typical open spina bifida cystica (Figure 3), though in some, the plaque may be covered by a thin membrane. In most instances a myelocele is associated with an Arnold-Chiari malformation of the hind brain, which together with aqueduct anomalies account for the well-established hydrocephalus present at birth in about 80% of cases. The latter is often rapidly progressive and in its control often presents major problems.[3] Interruption of the spinal pathways by the spinal lesion leads to variable paralysis of the legs, incontinence of urine and feces, anesthesia of the skin, and abnormalities of the hips, knees, and feet. Infants with myelocele have a poor prognosis and if left unoperated, 90% will die usually within 6 months of ascending meningitis or progressive hydrocephalus. However, if the open spinal lesion is operated upon promptly and progressive hydrocephalus is shunted, over 50% of infants will survive, but generally with severe physical and variable amounts of mental handicap.[4-6] In some cases the myelocele is covered by a thick membrane or full thickness skin (Figure 4), when the prognosis for survival is better as the risk of ascending meningitis is largely eliminated.

About 1 case in 20 of spina bifida cystica is a true meningocele (Figure 5) where the spinal cord itself is largely intact and the deficit is confined to the skeletal tissues allowing the meninges, generally covered by a thick membrane or occasionally full thickness skin, to protrude. Such infants have little or no neurological deficit and removal of the lesion leaves a more or less normal child.[1]

Complicated spina bifida occulta which usually involves more than one neural arch and is often associated with widening of the spinal canal, vertebral body abnormalities, a hairy mole or nevus, some abnormality of legs or feet, or some neurological deficit, must be regarded as part of the NTD spectrum of malformations.[7]

Simple spina bifida occulta, a purely radiological finding, where as a rule there is involvement of one neural arch only and no other skeletal or skin abnormality and no neurological deficit,and is found in perhaps 10% of the adult population is not part of the NTD spectrum of malformations and must be regarded as part of normal variation.[7] Hydrocephalus

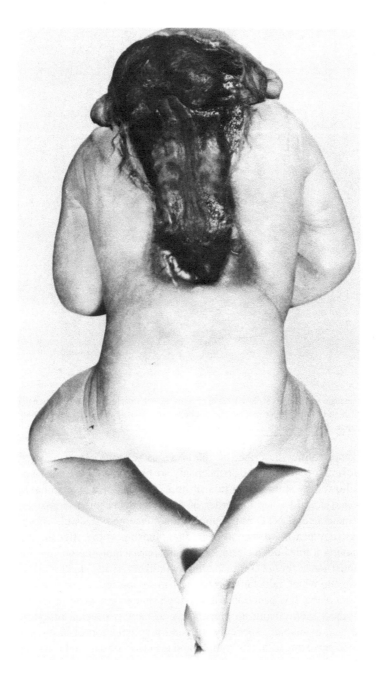

FIGURE 1.   A full-term anencephalic stillborn with partial spinal rachyschisis.

not associated with spina bifida seems also not to be part of the NTD,[8,9] nor are those rare conditions where an NTD is part of a malformation complex such as in Meckel-Grüber syndrome.

III. ETIOLOGICAL MECHANISMS

A. Genetic Aspects

Large family and other genetic studies[8,10,11] of the NTDs carried out in the U.K. have all shown the same family aggregation of cases and pattern of recurrence with any of the

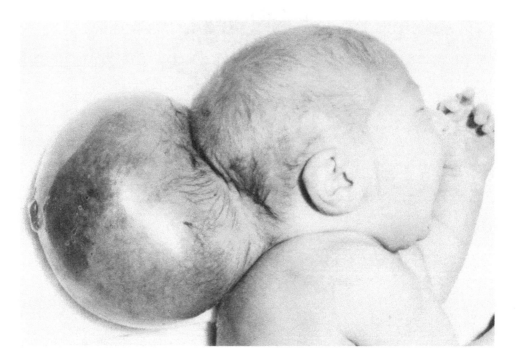

FIGURE 2.   A neonate with a large "closed" skin covered encephalocele which at operation was found to contain a large portion of both occipital lobes most of the cerebellum and part of the ventricles protruding through a wide defect in the occiput. The head itself is almost microcephalic.

spectrum of abnormalities described in the previous section, occurring in a family group. There is however, a tendency for anencephaly to be followed by anencephaly and spina bifida to be followed by spina bifida in a sibship. The risk of a woman having a recurrence once she has had a child is on the order of 5% though the precise risk depends to some extent on the incidence of NTD in the population concerned[12] as well as other factors. The risk for a pregnancy bearing a second degree relationship to the affected case is rather less and that for having a third degree relationship is approaching the population prevelance.

Recessive inheritance for NTD is made highly unlikely by the small proportion of children affected in a sibship which falls far short of the 1 in 4 expected. Dominant inheritance would assume an unacceptably low penetrance and cytoplasmic inheritance[13] is rendered improbable as affected mothers seem to run no greater risk of having affected offspring than affected fathers.[14] The data available, especially the family aggregation of cases, ethnic difference in the prevalance persisting after emigration, and the effect of parental consanguinity, suggest that the NTDs fit into a pattern of inheritance similar to that of the other common malformations. It is therefore suggested that NTDs have a multifactorial etiology resting on a genetic predisposition to develop the malformation which is polygenic depending on minor additive genetic variations at several gene loci, leading to a threshold beyond which fetuses are at risk of developing NTD if environmental trigger mechanisms act during the teratogenic period.[15] In this concept only a relatively small proportion of the general population of fetuses would be beyond this risk threshold (Figure 6). The various neural tube defects are therefore interrelated having the same etiological basis and representing different end products of the same general process.[12] Which one of the spectrum of abnormalities develops in any particular instance probably depends in part at least on the precise timing and the severity of the insult.

Experimental evidence suggests that there are almost certainly a number of environmental

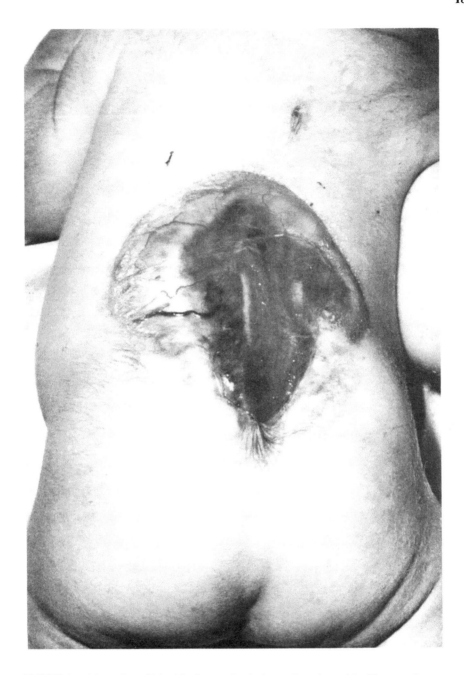

FIGURE 3. A large "open" dorsi-lumbar myelocele (or myelomeningocele) with a central zona vasculosa (neural plate), covered by a thin membrane merging periferally with the surrounding skin under which there is an accumulation of cerebrospinal fluid. This infant had paralysis of legs and sphincters and well-established hydrocephalus.

trigger factors responsible for these abnormalities in man. As there is little likelihood of being able to modify the polygenic component, primary prevention would therefore be dependent upon being able to identify these environmental trigger factors so that they can either be removed from the environment or be avoided.[16]

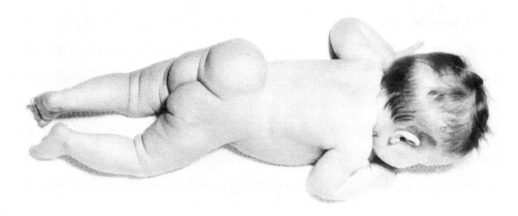

FIGURE 4.   An infant with a large "closed" skin-covered lumbro-sacral myelocele. Such infants have a better chance of survival because ascending infection is less of a problem, but often suffer from severe limb and sphincter paralysis.

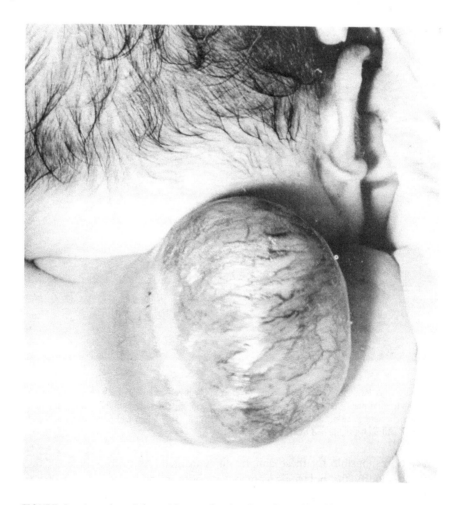

FIGURE 5.   A newborn infant with a cervico-dorsal meningocele, with a sac covered by thick membrane. There was no cord involvement and once the sac was removed, the infant developed normally without paralysis or hydrocephalus.

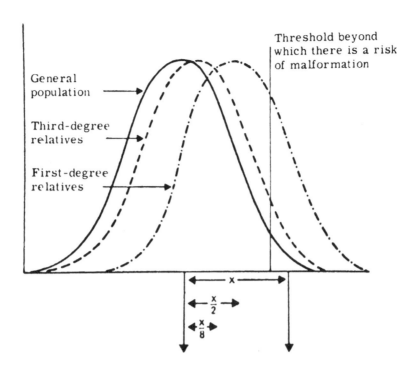

FIGURE 6. Model for polygenic inheritance of NTD. Only a small proportion of the general population of embryos would be beyond the threshold of risk. First degree relatives, whether sibs or children with half their genes in common with the index case have a distribution of the index case and the general population bringing many more beyond the threshold. X = Deviation of mean of malformed individuals from the population mean.[15]

B. Environmental Aspects

A great variety of environmental insults can greatly affect the growth and the development of fetuses in experimental animals. It was maternal vitamin A deficiency in pigs reported by Hale in 1933 which induced piglets without eyeballs and other defects[17] that first started modern teratology. To produce malformations experimentally the timing of the environmental insult is critical. As a rule, no malformations are produced if the insult occurs before implantation of the embryo has occurred or after the main organogenic processes have been completed. The teratogenic period is generally confined to the relatively limited time when organogenesis takes place, which for the neural tube in rodents is from around day 8 to day $11^{1}/_{2}$ after fertilization.

Maternal vitamin A deficiency, though causing other malformations, has so far not produced NTD in any species of experimental animal. Vitamin A excess, on the other hand, will regularly precipitate NTD in the mouse,[18] especially the curly-tailed mouse,[19] the guinea pig,[18] the rabbit,[18] the hamster,[20] and the pig.[21] Induced folic acid deficiency, using folic acid antagonists, will precipitate exencephaly and spina bifida in rats, mice, and cats.[22,23] Vitamin E deficiency induces exencephaly and hydrocephalus in rats but not in mice.[24] Pyridoxine (vitamin B_6) and vitamin B_{12} deficiencies also precipitate exencephaly in rats.[26,27] However, neither vitamin D[22] nor riboflavin lack[28] seem to lead to NTD and dietary vitamin C lack in experimental animals does not cause malformation, as most animals seem to synthesize their own. Only primates, including man and the guinea pig, require dietary ascorbic acid (AA). Only hypervitaminosis A produces NTD in laboratory animals and excessive doses of other vitamins such as vitamin D, vitamin E, and 6-amino-nicotinamide have induced malformation in other systems but not NTD.[29-31] Large quantities of folic acid seem to be without teratogenic effect.

It is generally recognized that great care has to be taken in translating experimental results to the human situation, and until quite recently no actual vitamin deficiency has been causally related to NTD in man. There has been one report recently where prolonged vitamin A dosage is said to have caused anencephaly[32] and another where it has been said to have caused urogenital abnormalities.[33]

Although a number of drugs, and for that matter also virus infections, have from time to time come under suspicion in causing NTD,[34] the only drug which has been shown to actually precipitate NTD in man is aminopterin, a powerful antifolic acid agent used at one time as an abortifacient.[35] However, on epidemiological grounds poor maternal nutrition as an etiological factor has been suspected for some time and there are pointers that this may be due to relative vitamin lack or more specifically lack of folic acid for the developing embryo. The much lower prevalence and the difference in character of the NTD seen in some of the mongoloid, and to a lesser extent the negroid populations than those of caucasian background can probably be explained on genetic grounds, especially as most of the differences are largely retained after emigration and intermarriage.[15,34] The great differences in the prevalance of NTD amongst populations of caucasian origin living in different parts of the world must in part be due to differences in environment and possibly nutrition. The Irish, with their high prevalence of NTD emigrating to North America, and the Jews in the Middle East moving to Israel were found to have a decreasing risk of NTD with each succeeding generation and eventually approaching that of the local population,[36,37] presumably because of progressive changes in their living standards and nutritional habits.

In the British Isles the more socially deprived regions such as the Irish Republic, Northern Ireland, South Wales, northwest England, and southwest Scotland, have the highest prevalence while those people living in the southeastern part of England and East Anglia have the lowest prevalence.[15] Within these areas, as well as in the British Isles generally, social class IV and V families, where the husband is a semiskilled or unskilled worker, have a much higher risk of NTD than the social class I and II families where the husband is in a professional or higher clerical occupation.[15,34,38] Undoubtedly the nutritional standards in the poorer parts of the British Isles are lower than in the more prosperous areas. Social class IV and V mothers are less well nourished than those from social class I and II.[38,40-42] The former buy fewer fresh vegetables and fruit; fresh green vegetables tend also to be less readily available in the winter than in the summer months. There has been a seasonal variation too with winter and spring conceptions ending more commonly in an offspring with NTD than conceptions occurring during the rest of the year.[34,38] These latter differences are probably also related to nutrition though at one time virus infection had been suspected.[39]

Considerable secular variations in the incidence of NTDs have occurred amongst the caucasian population in various parts of the world, including Britain, which have often been difficult to explain. It is tempting to attribute the sharp rise in the incidence of NTD in Boston during the depression years,[42] in Holland toward the end of World War II,[43] and in a number of German cities immediately after that war[44] to poor maternal nutrition seen at that time. There have been some variations in their prevalence and for that matter in the prevalence of both anencephaly and of spina bifida separately in the last 40 years which have so far defied explanation. However, since 1972 there has been a steady, general fall in the prevalence in Britain, which has been most marked in those regions where the NTD problem is the greatest.[45,46] In addition, there seems to have been a lessening of the social class differences and the seasonal variations. It would be nice if an explanation for these changes were to be found in an improvement in the general standard of living and of nutrition in Britain but there does not seem to be any firm evidence to support this.[47] However, it may be worth noting that in the last decade or so there has been an increasing availability of fresh imported green vegetables and an increased use of fresh, deep frozen foods, especially green food, which are freely available throughout the year.

Some caution has to be exercised in drawing conclusions from slight differences or variations in the prevalence or demographic characteristics of NTD reported at birth as slight variation in the quality of reporting NTD as well as slight differences in the incidence of miscarriage could greatly affect the figures. It has recently been found that even in the low prevalence populations a high proportion of early abortuses have anencephaly or spina bifida.[48]

IV. DIETARY AND SUPPLEMENTATION STUDIES

Accurate studies of the diet taken are difficult to carry out and most of the methods used are open to criticism. The 24-hr recall method may accurately record the intake of the previous day which may, however, not have been typical of the usual diet. The weighing method which is generally reckoned to require the accurate weighing of the dietary intake over a number of days, usually a week, is both cumbersome and expensive. It is difficult in unintelligent subjects and often requires professional supervision. The very act of scrutiny of the diet in this way is likely to cause it to be modified. Finally, there is the method of recording the pattern of meals taken over the week or a longer period,[49] without actually recording the amounts of the individual food items employed in our studies.

There have been few reports of studies of the diet taken in early pregnancy. Smithells and his team[50] reported the results of an extensive study into the diet during the first trimester of almost 200 women and found that the intake of all nutrients was significantly lower in women from social class III, IV, and V than those from social class I and II. This was paralleled by blood levels of folic acid and other micronutrients in a study of about 900 first trimester pregnancies.[51]

A. South Wales Studies
1. First South Wales Dietary Study (1969—1975)
With the epidemiological data possibly suggesting a dietary etiology, the report of the effect of aminopterin in precipitating NTDs in man, the animal experimental evidence, and some information on folic acid metabolism abnormality in some women who had a child with an NTD,[52] a folic acid supplementation study was undertaken to try to reduce the risk of recurrence of NTD in women at increased risk. However, it was felt that it was unrealistic to do so without also looking both retrospectively and prospectively at the diet of these women and study the possible beneficial effects of dietary counseling.[53]

Nine hundred and two (902) women residents of Glamorgan and Gwent who were all under the age of 35 and therefore more likely to plan to have further children, and who had had a pregnancy ending in an NTD between 1954 and 1974, were ascertained from a variety of sources. Two medically qualified field workers experienced in diet taking visited them in their homes, 425 women being visited in the eastern half of the area by the one and 477 in the western half by the other. There were no differences in the incidence, the social class structure, or the dietary ratings in the two areas. A questionnaire was administered concentrating on the social, medical, obstetric, and family histories together with a simple dietary history showing the usual pattern of meals when not pregnant and variations during the first trimester of each of their pregnancies. A check list of foods was also used which was designed to reveal any lack of intake of foods rich in folic acid. Pregnancies were divided into those which ended in an NTD (index pregnancies) and all the others. A sample of blood was taken for hematological investigation and for serum and red cell folate and vitamin B_{12} estimation using the *Lactobacillus casei* and the Euglena methods, respectively.[55,56] Blood samples were taken only from the 443 women who felt that they had not completed their families. These were asked to cooperate in a prospective study of their next pregnancies. In the western area only, all women were given dietary counseling and advised to take an

Table 1
FIRST DIETARY STUDY:
QUALITY OF DIET DURING INTERPREGNANCY
PERIOD COMPARED TO THAT DURING PREVIOUS
PREGNANCIES

	Quality of diet						All women (n)
	Good		Fair		Poor		
Period of diet	n	%	n	%	n	%	
Interpregnancy	24	14	106	61	44	25	174
Pregnancy with NTD	17	10	64	37	32	53	174
Other pregnancy	23	19	63	51	37	30	123

Note: The interpregnancy diet was the diet currently consumed at the time of the first survey; the pregnancy diets related to the first trimester of previous pregnancies and were obtained by recall.

optimum diet at the latest from the day they discontinued any form of contraception until their baby was born. No counseling was given to the women in the eastern area. All women were asked to notify the investigators immediately that a menstrual period was overdue, but not later than 8 weeks after their last menstrual period. They were visited again as soon as possible afterward. The dietary history was updated and another specimen of blood was taken for laboratory investigation. Further visits were paid at the end of the first and second trimesters and after the baby's birth to check the diet and to take a further sample of blood.

In analyzing the diets, each essential nutrient was rated separately, the mean general rating being applied to the quality of the diet as a whole. Any excess of refined carbohydrate was considered to imbalance the diet overall. Ratings used were good, fair, and poor. A good diet was one with daily helpings of foods containing protein, vitamins, folic acid, calcium, and iron, without an excessive intake of confectionary and soft drinks. Diets deficient in the essential nutrients where the calories were made up largely by refined carbohydrate foods and fats were rated poor. Fair diets were intermediate in quality where some nutrients might be rated good and others poor or the total balance might be affected by large amounts of refined carbohydrate and fats.

Altogether, 174 women who volunteered and complied achieved 186 pregnancies; by the end of 1974, 103 women in the western area who received dietary counseling had 109 pregnancies, and 71 in the eastern part had 77. Twelve women each had two pregnancies by the end of 1974. It soon became apparent that in the past only a minority of these women consumed a good diet. The poorest diets seemed to have been taken during the first trimester of the pregnancy ending in the NTD, partly due to severe anorexia and vomiting, a common feature of those affected pregnancies. Diets during the first trimester ending in a normal baby were usually not as poor as those during the first trimester of pregnancies ending in an NTD, but still not as good as during the interpregnancy period (Table 1).

Though it was known whether the mother came from the area where they would have been given dietary counseling, all the diets taken during the first trimester of the prospective study pregnancies were assessed before the outcome of the pregnancy was known and checked later without reference to the outcome. There was virtually no discrepancy between the two assessments.

In West Glamorgan where all women had been counseled to improve their diet, the majority heeded this advice and almost three fourths (71%) had improved their diet in time for their

Table 2
FIRST DIETARY STUDY:
CHANGE IN DIET IN PROJECT PREGNANCY COMPARED
TO THAT IN INDEX PREGNANCY

	Change in diet						All pregnancies (n)
	Improved		No change		Worse		
Instructions on diet	n	%	n	%	n	%	
Counseled	78	71	29	27	2	2	109
Not counseled	9	12	63	82	5	6	77
All diets	87	47	92	49	7	4	186

Note: (1) Six counseled women and six noncounseled women had two pregnancies; (2) the project pregnancy diet was recorded at the time of this pregnancy; and (3) the index pregnancy diet was taken by recall at the interpregnancy interview.

Table 3
FIRST DIETARY STUDY: OUTCOME OF
PROJECT PREGNANCIES BY QUALITY OF
DIET IN FIRST TRIMESTER

Outcome of pregnancy	Quality of diet			All pregnancies (n)
	Good (n)	Fair (n)	Poor (n)	
Normal child	53	85	22	160
Recurrent NTD	—	—	8	8
Miscarriage	—	3	15	18
All outcomes	53	88	45	186

next pregnancy. The reason that many of the remainder did not do so was frequently because of excessive nausea and vomiting (Table 2). In East Glamorgan over 80% did not change the pattern of their diet. The outcome of the 109 pregnancies in West Glamorgan was 10 miscarriages and 3 recurrences of NTD; amongst the 77 in East Glamorgan there were 8 miscarriages and 5 recurrences. The ratio of the risk of recurrence in East Glamorgan compared to West Glamorgan was 2.4, although this did not differ statistically significantly from 1.0 (Table 3).

All eight recurrences, however, occurred in women whose diets were considered to be poor during the first trimester of pregnancy, an assessment which was made at the time of the first visit during the pregnancy before any outcome could be known. There were no recurrences in the pregnancies with a good or fair diet. This distribution was most unlikely to have occurred by chance ($p < 0.001$).

Of the three women with recurrences who received counseling, one had ignored the advice but two suffered nausea and vomiting during the first month after conception which had continued into hyperemesis gravidarum.

2. Second South Wales Dietary Study (1974—1979)

It was felt that the diet needed more detailed investigation and that two control groups should be included. The same area was covered by the same two medical field workers. Over 200 (244) women who had an NTD from 1964 onward, including 36 women who were also in the previous study, were the index cases. Control groups consisted firstly of 123 sisters of these index cases who had only normal children and there was no essential difference in the social class between the index women and their sisters as judged by the husband's occupation. The second control group were 50 university and professional wives. The same questionnaire was used for the social, medical, obstetric, and family histories as in the first study, but this time the dietary history was much more detailed and was based on Burke's method. This method gives due consideration to economic and social background and food customs, spanning a lifetime's eating habits, including childhood, before and after marriage, and during each pregnancy.[49] In addition, a checklist of foods was again used. All index and control women received dietary counseling if they were planning to have another child. They were again asked to notify us immediately when they thought they were pregnant and were re-interviewed as soon as possible to check on the diet. The interview for this second study took between 2 and 3 hr to complete, whereas for the first dietary study, the interview was usually much shorter, even in the western part where dietary counseling was given.

Diet ratings made before the outcome of prenatal diagnostic tests or the outcome of prospectively studied pregnancies were known were again good, fair, and poor (poor a), but this time a fourth category was included where the diet was fairly good but was unbalanced by excess of refined carbohydrates and fat (poor b).

A much clearer picture of the eating habits of women emerged in this study, as they were encouraged to describe in their own words their pattern of eating, often acquired in the parental home and continued after marriage. This may have been modified if they were able to afford more but the additional income was generally spent on refined carbohydrates and convenience foods rather than on improving the intake of essential nutrients.

The following are definitions characteristic of some of the women on inadequate diets:

> **Pickers,** who are women who toy with their food and rarely eat more than a fraction of the helpings on their plate.
>
> **Nibblers** tend to eat incessantly, especially when viewing television or feeling lonely, bored, or anxious. Some of these eat between regular meals, others never sit down to a proper meal, although they may pick at the dinners prepared for their husband's homecoming.
>
> **Gorgers** are those who overeat at meals and those who are also "nibblers" become overweight. This leads to slimming fads and crash diets, interspersed with lapses into compulsive eating. Some of the more figure-conscious then starve themselves for weeks at a time and if this coincided with conception their diet would be markedly deficient in all the essential nutrients during the vital period for the development of the embryo's nervous system.
>
> **Moody women** are those who also tend to be erratic in their diet and when depressed either go off their food or eat more than usual in an attempt to comfort themselves.
>
> **Food faddists** are those who have an aversion to one or more foods often including several essentials. A number of these women said they disliked meat, fish, eggs, greens, fruit, and salads so they subsisted almost entirely on refined carbohydrates and fats.
>
> **Chip maniacs** was a term often used by their sisters to describe an index mother who ate mounds of chips (french fried potatoes) for one or two meals daily, sometimes with chip sandwiches in between and very little other foods.

Many index women who described themselves as always having had a "sweet tooth", ate a surprising amount of refined carbohydrates. In one average week, such women consumed as much or even more than 2 lbs of sweets, 7 bars of chocolate, 10 packets of crisps

Table 4
SECOND DIETARY STUDY:
INTERPREGNANCY DIET — CASES AND CONTROLS
NUMBER OF SUBJECTS AND ROW PERCENT

	Quality of diet								
	Good		Fair		Poor (a)		Poor (b)		All
Group of women	n	%	n	%	n	%	n	%	cases
Index cases	34	14	110	45	33	14	67	27	244
Sisters controls	43	35	48	39	23	19	9	7	123
Upper class controls	39	78	8	16	3	6	0	0	50

Note: (1) All diets were current interpregnancy diets; (2) index cases had a pregnancy with a fetal NTD; (3) sisters of index cases who had not had a NTD; (4) upper class controls, university, and professional wives who had not had a NTD; and (5) poor diets are subdivided into (a) those with an adequate diet but excess carbohydrates and fat and (b) those with a totally inadequate diet.

(potato chips), 7 whole packets of biscuits, and several bottles of ''pop'' (carbonated soda), as well as two or three spoonfuls of sugar in the many cups of tea and coffee that they would drink. Their sibling controls often commented with disapproval that their index sisters were ''living out of tins'' (i.e. tinned meats, peas, baked beans, fruit, milk, and cream) or, that they ''couldn't be bothered'' to cook fresh meat and vegetables but lived on convenience foods, chips, and large amounts of refined carbohydrate foods.

In contrast, the upper class controls, although often on a tight budget, were more interested in cooking, enjoyed a more varied diet and recognized the importance of foods containing protein and vitamins, especially during pregnancy.

The quality of the usual interpregnancy diet in the index cases showed almost the same ratings as in the first study with only 14% rated as good and 2 out of every 5 on poor diet (Table 4). This compares with only just over one in four of the sibling controls being on poor diet. Four out of five of the upper class control women were on a good diet but some 6% were in fact taking a diet which was deranged by excessive amounts of carbohydrates and fats, very often in the form of sweets and chocolates, cakes and biscuits, soft drinks, and numerous cups of sweetened tea. The diet during the first trimester of the 251 pregnancies ending in an NTD in the index women had been worse than during the first trimester of their 152 other pregnancies. Alternately, in the 176 further pregnancies reported by the end of 1979, they had considerably improved the quality of the diet during the first trimester with only 18% being poor, confirming again that dietary counseling is followed by an improvement in the diet of a majority of women. There were five recurrences, again all in women who remained on an inadequate diet (Table 5).

B. Preconceptional Folic Acid Supplementation Trial (1969—1975)

A double-blind, randomized placebo controlled trial of folate treatment was therefore mounted simultaneously with the first dietary study. The same 443 women were used (Glamorgan and Gwent) who had previously had a pregnancy ending in an NTD and who felt they had not yet completed their families. Dietary inquiries were made as detailed in the previous section. With no promises of a possible beneficial effect they were asked to volunteer to take either tablet A or B, only one of which contained folic acid, from the day they discontinued contraception until the end of the first trimester of their next pregnancy.[54,59,85] Neither the women nor the field workers knew the contents of the tablets allocated but those allocated to folic acid received 4 mg/day. A blood sample was taken for

Table 5
SECOND DIETARY STUDY:
DIET OF INDEX WOMEN DURING FIRST TRIMESTER OF
PROJECT PREGNANCY

| | Quality of diet | | | | | | | | |
| | Good | | Fair | | Poor (a) | | Poor (b) | | All pregnancies |
Outcome to fetus	n	%	n	%	n	%	n	%	(n)
Normal	68	40	76	44	15	9	12	7	171
NTD					3	60	2	40	5
All fetus	68	39	76	43	18	10	14	8	176

Note: (1) Seven index women had had two previous pregnancies with NTD; (2) 176 index women became pregnant and 5 had a NTD giving a rate of 2.8% (all recurrences had fair or poor diets); and (3) All women were counseled to improve their diet.

hematological investigation and folate and vitamin B_{12} estimations. They were asked to notify investigators of a missed menstrual cycle or period as soon as possible, but at least within 8 weeks of the last missed period (LMP). They were then visited soon after when a further sample of blood was taken for folate and other estimations. They were revisited at 3 months, at 6 months, and after the baby's birth.

Following the "Thalidomide disaster" the climate of opinion regarding medication in pregnancy was generally unfavorable; consequently about one third of the women refused to join the folic acid supplementation trial though they were quite willing to cooperate in the diet study. Some became pregnant sooner than planned and so were not supplemented early enough and others failed to notify investigators of the pregnancy early enough and some did not have the necessary blood folate estimation available. In this event only 123 women who volunteered for the supplementation study achieved a pregnancy by the end of 1974 and also apparently complied with the instructions, etc. Of these, 60 turned out to have been allocated to folate supplementation and 63 to placebo. Since ingestion of folic acid is followed by a sharp and considerable rise in the serum folate level, and rather later by an elevation of the red cell folic acid level as well, any woman whose serum contained less than 10 $\mu g/\ell$ of folic acid was assumed not to have taken the folic acid tablets and was classified as a noncomplier (Figures 7 and 8). By this criterion 15 women failed to take the folic acid tablets and only 45 were supplemented. With several of these noncompliers, although the women themselves agreed to participate, it appeared to be the husbands who later objected to the taking of any tablets in pregnancy that were not strictly necessary.[60] The women concerned sometimes seemed reluctant to admit to the change of mind. There was one woman in the "supplemented" group with a recurrence of an NTD and there were five recurrences amongst the nonsupplemented (those on placebo and the noncompliers) (Table 6). As this was not statistically significant the results of the study were not published in 1975. However, in 1980 the one apparently supplemented woman with a recurrence confessed to not having taken the tablets as instructed but in large numbers only just before she was revisited by the field worker after reporting her pregnancy at 8 weeks, so accounting for the very high serum level of folate.[59] As she was not supplemented during the teratogenic period she too was now classified as a noncomplier. Thus, the 44 supplemented women, including 10 who were on an unsatisfactory diet had no recurrences, while the 79 unsupplemented had 6 recurrences (Table 7). This now was a statistically significant difference ($p > 0.04$ Fishers exact test with a single tail), suggesting that folic acid supplementation seemed effective in preventing recurrences of NTD.

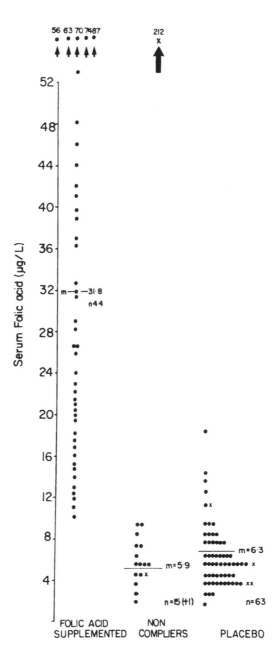

FIGURE 7. Serum folic acid levels during the first 123 pregnancies included in the folic acid intervention trial divided up into the supplemented, the placebo and the noncomplier group. The recurrences are indicated by crosses. With the exception of the one noncomplier who took the large number of folic acid tablets at 7 weeks (who has been excluded from the calculation of the means) the serum folate levels of the noncomplier and the placebo groups are almost identical.

The specific effect of folate has to be separated from the nonspecific effect of diet. There were no recurrences among the 91 women who received good or fair diets, but there were 6 recurrences among the 32 women receiving a poor diet ($p < 0.001$, Fishers exact test). As we have shown in the South Wales dietary study, women who take poor diets are at an

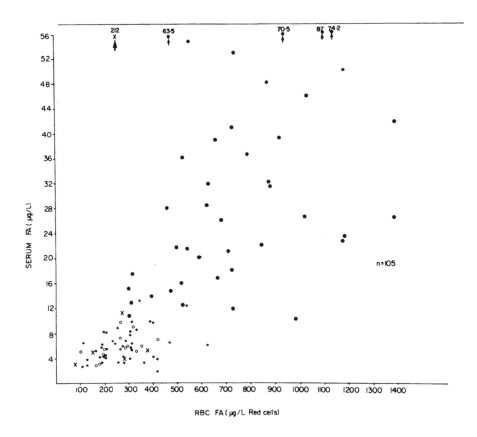

FIGURE 8.   Serum and red cell folate levels during the first trimester in the 123 pregnancies included in the folic acid intervention trial. All the recurrences (X) the placebo (•) or noncompliers (o), with the exception of one woman who took a large amount of folic acid at 7 weeks, have both low serum and red cell folate levels. The supplemented group (●) have elevated folate levels.

extremely high risk of recurrence of fetal NTD. Within this high-risk group of women, however, there were no recurrences in the 10 who had taken folate supplementation but 6 recurrences in the 22 who had not taken supplementation (Table 8). Thus, although there may have been some bias owing to women who were receiving an inadequate diet also failing to comply, within this group the preventive effect could still be detected.

What has been demonstrated is that in the circumstances of South Wales in 1969—1979 in the three investigations it was very difficult to establish a trial of preventive treatment in pregnancy and there was no assurance that the subjects would actually comply with a protocol. The two dietary studies were based upon a pragmatic classification of the quality of the diets, but the detailed histories on which they were based were extensive and had been taken by two clinical field workers widely experienced in taking dietary histories from women in these population groups. They demonstrate first that women who have had a fetal NTD are still currently taking a poor diet, that it is possible to persuade women to change their diet (three fourths of the counseled group made considerable improvements compared to only about one tenth of the group not counseled), and that in the prospective survey all the recurrences occurred in the group of women who were shown to have a deficient diet from the history taken contemporaneously in the second month of pregnancy.

The supplementation trial demonstrates first that such trials are very difficult to undertake. We offered no assurances, placed no pressure on the women to join the trial, avoided publicity, and did not follow up those women who did not return voluntarily to the field

Table 6
FOLIC ACID SUPPLEMENTATION TRIAL: RECURRENCE OF NEURAL TUBE DEFECTS

Case	Treatment	Diet	Blood estimations				Outcome		
			Week	Serum folate	Red cell folate	B_{12}	Delivery	Lesion	Gestation mos.
A	FNC	Poor	7	212.0	259	112	MC	AN	3
B	FNC		8	4.8	155	167	LB	SB	9
C	PL		8	3.7	228	89	LB	SB	8
D	PL		7	11.0	275	210	MC	AN	4
E	PL		8	5.0	380	217	LB	SB	9
F	PL		7	2.9	76	371	T	AN	5

Case A: Past history: 1 surviving spina bifida, 1 miscarriage; diet known to be poor; excessive carbohydrate intake; took no tablets until 7 weeks after LMP; she then took a considerable number 24 hr before blood sample was taken and then continued to take the tablets as instructed; we did not know this until after the end of the whole trial; outcome: anencephalic miscarriage at 12 weeks.

Case B: Past history: 2 normal children, 2 anencephalics, and 4 spontaneous abortions; diet very poor; started to take tablets 7 to 10 days after LMP at a time when she was feeling sick; off work with febrile illness; the nausea increased to vomiting and she returned most of the tablets she was taking; her diet up to 8/52 after LMP was largely milk; thereafter, she felt much better; outcome: full-term spina bifida female, died on first day.

Note: Estimations in μg/ℓ. Abbreviations: FNC = folate noncomplier; PL = placebo; LB = live birth; MC = miscarriage; AN = anencephalic; SB = spina bifida cystica; T = termination.

Table 7
FOLIC ACID SUPPLEMENTATION TRIAL:
OUTCOME OF PREGNANCY BY TREATMENT GROUP

Outcome of pregnancy	Folate supplement		Placebo supplement	All cases
	Compliers	Noncompliers		
Normal fetus	44	14	59	117
Neural tube defect	0	2	4	6
All outcomes	44	16	63	123

Table 8
FOLIC ACID SUPPLEMENT TRIAL:
OUTCOME OF PREGNANCY BY QUALITY OF DIET IN
FIRST TRIMESTER BY FOLATE SUPPLEMENTATION
RECEIVED

Outcome of pregnancy	Quality of diet							
	Good		Fair		Poor		All diets	
	F	NF	F	NF	F	NF	F	NF
Normal fetus	18	25	16	32	10	16	44	73
Neural tube defect	0	0	0	0	0	6	0	6
All outcomes	18	25	16	32	10	22	44	79

Note: F = Folate supplemented.

workers. There is a serious problem in how to ensure that any treatment has been taken and how this can be validated.

It is essential to distinguish between the effectiveness of an agent which has been used in preventing recurrences and the efficiency of a regime in converting an effective agent into a usable efficient system of treatment. The trial demonstrated that no recurrences occurred to women who took the folate, whether or not they had an adequate diet; similarly no recurrences occurred to women who had an adequate diet whether or not they were supplemented. All the recurrences occurred in the small group who had poor diets and were not supplemented. If the serum folate level and the red cell folate could be raised above critical thresholds, then there were no recurrences of NTD in a high risk group of women.

The trial therefore seems persuasive that NTDs are the result of deficient diet and that they can be prevented either by an adequate diet or by supplementation with folate. It does not exclude the possibility that there are other deficiencies or some abnormality of folic acid metabolism which can be swamped by an excessive intake of folate.

C. Periconceptional Supplementation with Pregnavite® Forte F

The finding that various nutrients are lower in the diet of women from social class III, IV, and V than from social class I and II,[50] and more particularly, the finding that red cell folic acid, riboflavin, leukocyte vitamin C, and serum vitamin A levels are also significantly lower in the lower social class mothers led Smithells and his colleagues to the conclusion that sub-clinical deficiencies of one or more vitamins contribute to the causation of NTD. This was reinforced by their report of particularly low levels of these micronutrients in the first trimester of pregnancies complicated by a fetus with an NTD.[51] They tried to launch a placebo-controlled multivitamin supplementation trial, but were not able to obtain ethical permission for the use of placebo, therefore they had to work with less satisfactory controls. Their index patients, recruited mostly from the genetic counseling centers of Yorkshire, Lancashire, Cheshire, Northern Ireland, and the southeastern part of England, were volunteer women who have had one or more pregnancies ending in a fetus or infant with an NTD and who were planning further pregnancies, but were not pregnant at the time. They were asked to take three Pregnavite® Forte F tablets per day, not less than 28 days before conception, and to continue taking these until the date of the second missed period. Control women were from the same areas, especially Northern Ireland. They were already pregnant at the time of recruitment or they declined to take the tablets.[61,62] Each tablet of Pregnavite® Forte F (Bencard) contained vitamin A (4000 I.U.), vitamin D (400 I.U.), thiamin (1.5 mg), riboflavin (1.5 mg), pyridoxin (1 mg), nicotinomide (15 mg), ascorbic acid (40 mg), folic acid (0.36 mg), ferrosulfate (equivalent to 75.6 mg Fe), and calcium phosphate (480 mg).

Over 100 (178) supplemented pregnancies apparently resulted in 140 infants without NTD, one infant with an NTD, and 11 spontaneous abortions which were examined and found to have no NTD. Twenty-six (26) pregnancies were still continuing but with a normal amniotic AFP level found at amniocentesis in mid-trimester. Two hundred and sixty (260) control pregnancies resulted in 192 infants without NTD, 12 infants or fetuses with an NTD, and 17 spontaneous abortions, one of which was found to be of a fetus with an NTD. Thirty-eight (38) pregnancies were still continuing but with a normal amniotic AFP level after amniocentesis (Table 9).

Smithells and his colleagues[63] added a second cohort to these already significant results which altogether gave three recurrences in 397 Pregnavite® Forte F-supplemented pregnancies and 23 recurrences in 493 unsupplemented control mothers ($p < 0.0003$, using Fishers exact test).

Table 9
PREGNAVITE® FORTE F
SUPPLEMENTATION TRIAL[61]

	Supplemented	Controls
Infant/fetuses with NTD	1	12
Infant/fetuses without NTD	140	192
Normal amniotic AFP	26	38
Spontaneous abortions		
NTD	0	1
No NTD	11	17
Total	178	260

V. DISCUSSION

Studies of diet, especially in pregnancy where dietary habits tend to change and become complicated by nausea and vomiting, are liable to be difficult. Retrospective inquiries into food intake are not only difficult but also likely to be unreliable especially if the time interval between the pregnancy and the inquiry is at all prolonged. The replies are also likely to be biased, as mothers may tend to exaggerate problems surrounding any pregnancy that resulted in an abnormal baby. Nonetheless, reasonably consistent results were obtained. Prospective studies of diet, such as the ones related in this chapter are less likely to be subject to the difficulties encountered in retrospective inquiries, especially if the assessment of the diet is made before the outcome of pregnancy is known. The dietary rating used in the study might be questioned. However, as there is quite a close relationship between the quality of the diet and the red cell folate level, the ratings made would appear to be valid ones. Those between the diet and the serum folate are less close, the latter being much more immediately influenced by the time of the last meal in relationship to the taking of the sample of blood (Table 10) and the content of that meal. However, in any one case, the red cell folate level alone need not be a direct guide to the quality of the general diet taken. There seems to be no relationship between the quality of the diet and vitamin B_{12} levels.

The importance of a good, balanced maternal diet cannot be overly stressed, not only for the sake of the general health of the community but also because of its relationship to NTD. Improvement in the maternal diet and its sensible preparation[64] following dietary counseling seems to reduce the risk of recurrence in women who have already had a pregnancy that ended in an NTD. It therefore follows that any such women should be given dietary counseling before they embark upon another pregnancy. Dietary counseling for motivated women as these usually are, carried out sympathetically and authoritatively need not be a time-consuming task, but it will need to be repeated for a later pregnancy as women tend to relapse into their former dietary habits. Preventing recurrences in women at increased risk will, however, only make a small impact on the NTD problem in the community; over 90% of all NTDs are born to women who have no obstetric or family history of NTD.[12] Therefore, to substantially reduce the number of NTDs in the community one must aim to generally improve the maternal diet through health education which should become an integral and important part of the school curriculum.[53] However, improvement in the diet of the population is likely to be slow because dietary habits founded in the parental home tend to be relatively resistant to change. There is, therefore, a need for another approach to prevention of NTD in the British Isles, at least in the immediate future, which may well be in the form of supplementation with folic acid or a multivitamin preparation containing folic acid.

With their findings that a number of micronutrients are low in the first trimester of

Table 10
FIRST DIETARY STUDY:
WOMEN WITH A PREVIOUS FETAL NTD

Quality of Interpregnancy Diet and Mean Concentration of Serum Folate
RBC Folate and Serum B_{12}

Quality of diet	Subjects		Mean serum folate ($\mu g/\ell$)	RBC folate ($\mu g/\ell$ of red cells)	Serum B_{12} ($\mu g/\ell$)
	n	%			
Good	65	16	8.9	295	283
Fair	197	47	5.4	236	272
Poor	149	36	4.9	197	285[a]
Not known	4	1			
All women	415	100	5.9	238	277

[a] Two sera with concentrations of B_{12} over 1000 g/ℓ excluded as subjects were likely to have been on B_{12} treatment; only those women who wanted further children who were not already pregnant on whom a satisfactory dietary rating and folate estimations were available are included in this table.

pregnancy in women from social class III, IV, and V,[50] and that red cell folic acid, leukocyte vitamin C,[51] and serum vitamin B_{12} levels[67] are low in the first trimester of pregnancies complicated by NTD, it is perhaps not surprising that Smithells and his colleagues chose a multivitamin preparation which contained 10 substances, including folate for their supplementation studies. There is no indication from their results as to whether the beneficial effect was obtained by all or some of the constituents of the Pregnavite® Forte F acting synergistically or whether the effect was due solely to the action of one of the substances, folic acid. Although the effect of poor nutrition on the developing embryo may be due to a deficiency in a number of micronutrients acting together and interfering with the orderly closure of the neural tube, or by allowing some as yet unidentified teratogen to have an influence, it is likely that the main factor in the British Isles is the lack of folate for the embryo at a critical stage of development. The evidence for this is epidemiological (this has been summarized in the section on environmental factors), experimental, embryological, and also clinical. NTD can be induced in the rat by exposing developing embryos to a variety of teratogens including antifolic acid agents after the eighth day following fertilization, the time when the neural tube begins to form; its closure is completed toward the middle of the twelfth day. This is also the stage in the rat embryo when the chorioallantoic placenta begins to develop and the embryonic heart starts to perfuse the placenta and the embryo.[68] From this, the pace of development and energy requirement increases significantly. During this time there are also considerable changes in the metabolic processes. Before the neural tube is formed the rat embryo derives its energy from glycogenesis and the pentose phosphate shunt, but from that time there is a progressive switch to the Krebs cycle. The Krebs cycle seems to be fully operational about the time that the neural tube is fully closed.[69,70] In the chick embryo it has been shown that neural tube closure is critically dependent on an oxygen gradient in the opposing neural crests, producing a narrow band of hypoxic surface cells in the opposing neural crests.[71] Lack of sufficient folate at that stage or a general metabolic upset could then interfere with the normal closure process; either the crests do not meet correctly or having met with the hypoxic necrosis that has progressed too far, dehisce again.

It may be thought that these findings in rodent or bird embryo bear little relation to the situation in man. However, it seems to be an accepted principle of comparative embryology that species differences in form and structure become less striking as earlier embryonic stages are examined and it is an interesting, but largely unstudied possibility that inter-species metabolic differences are also minimal in the early embryo.[72]

Translated to the human embryo, all these processes take place between the twenty-first and the twenty-fifth day after fertilization. There is every reason to suppose that they differ only in detail and in timing from those in the chick and the rat. It is not possible to investigate the metabolic processes in human embryos at that stage of development because the maternal circulation, the primitive placental barrier and tissues, and the fetal circulation intervene. However, further research using the whole embryo culture technique of New[73] should illuminate the precise changes in folate requirements and the influence that the various other micronutrients have on development. This may give some indication of the possible role that they play in early human development.

Clinical evidence for the suspicion of folic acid as etiologically important is no longer confined to the disastrous experience of the 1950s when aminopterin was used to induce therapeutic abortion. The chemical's use led to the birth of several children with NTD.[35] The low red cell folate levels in the first trimester of women from the lower social classes[50] and in pregnancies complicated by a fetus with an NTD[51] are indicators. As such the apparent beneficial effect of preconceptional supplementation with high doses of folic acid alone in preventing recurrences[54] in women at increased risk should not be overlooked.

Shortage of folic acid for the embryo need not be due to lack of folic acid in the diet, though some women undoubtedly subsist almost entirely on carbohydrates and fats and eat virtually no folate-containing foods. However, in many instances much of the folic acid that may have been present in the dietary items purchased are likely to be destroyed in previous storage and in processing such as canning or in the subsequent preparation of meals by for example, prolonged cooking.[57,64] In other instances there may be interference with absorption of adequate amounts of folic acid present in the meals. This may occur with the absorption of dietary zinc when the diet is unbalanced with excessive amounts of carbohydrate and phytate-containing foods and fats.[74,75] In any case zinc seems to be an essential factor for the enzymatic conversion of the polyglutamate form of folic acid present in vegetables to the monoglutamate form before intestinal absorption can take place.[76] Finally, some women might have a block or abnormality in their folic acid metabolism, as suggested by the finding that women who have just delivered a child with an NTD more frequently have a raised urinary excretion of formiminoglutamic acid (FIGLU) after histidine loading than do matched women delivered of a normal infant.[52] In some women oral contraceptives seem to interfere with folate metabolism.[77] It was because of a possible metabolic abnormality that a dose of ten times the estimated normal daily requirement of folic acid[78] was used in the supplementation study in order that any metabolic block might be overcome.

Contrary to what has been suggested recently,[79] the time has not yet come for preconceptional supplementation on a large scale. As the number of pregnancies in both the supplementation studies were relatively small, and as there were methodological criticisms, in the folic acid supplementation trial caused by the poor compliance[80] and in the Pregnavite® Forte F supplementation trial due to the poor choice of controls,[81] the results should first be confirmed with another larger and well-controlled study.[66] The Medical Research Council in London is planning to carry out just such a study by launching a large, multicenter, randomized, controlled trial to test the relative merits of folic acid, and a multivitamin preparation with and without folic acid, against a tablet containing only minerals. This will take at least 2 years to complete. If supplementation with folic acid is shown to be as effective as supplementation with multivitamins with folic acid, and is shown to be more effective than the multivitamin without folic acid or placebo, then the etiological importance of folate unavailability to the embryo in the British Isles will have been confirmed. Much more importantly, however, it will then enable folic acid (a cheap and safe substance costing less than £2 for 1000 days of supplementation) to be offered not only for the relatively few planned pregnancies of women known to be at increased risk for an NTD, but also for all planned pregnancies whether there is an increased risk or not. Thus, hopefully it may be possible to prevent a much larger proportion of the NTDs in the British Isles and perhaps in other parts of the world as well.[66] Ideally, folic acid should be added to some staple item of the diet which would also protect the many unplanned pregnancies. Because of the danger of precipitating subacute combined degeneration of the cord in marginally vitamin B_{12}-deficient elderly people, this may not be done safely. However, folic acid could quite safely be incorporated into the contraceptive pill or included in the "pack" to be taken on those days when the Pill is normally omitted.

If effective preconceptional supplementation with folic acid to prevent NTD is widely introduced, improvement in the general diet and especially in the maternal diet before conception must not be overlooked. However, it is likely that neither supplementation nor improvement in the maternal diet will be as effective in the low incidence parts of the world as it may well be in the British Isles. Finally, it must be remembered that there are almost certainly a number of environmental factors that trigger off NTDs. The search for further "triggers" should therefore continue.[66]

ACKNOWLEDGMENTS

Thanks are due to the editors and publishers of *Advances in the Study of Birth Defects*[82] for permission to use Figures 1 and 3; of *Fetal and Neonatal Pathology*[83] for Figures 2 and 5; and of *Birth Defects: Risks and Consequences*[84] for Figures 4 and 7. I am grateful for the help of both Dr. N. James and Dr. M. Miller for their tremendous efforts in carrying out the field work in the South Wales Studies and to Professor H. Campbell for his help in working out and evaluating the results.

The South Wales Dietary and Folic Acid Supplementation studies were supported by grants from Action Research for the Crippled Child and from Tenovus.

REFERENCES

1. **Laurence, K. M. and Tew, B. J.**, The natural history of spina bifida and cranium bifidum: malformations in South Wales. IV. *Arch. Dis. Child.*, 46, 127, 1971.
2. **Laurence, K. M. and Weeks, R.**, Abnormalities of the central nervous system in *Congenital Abnormalities in Infancy*, Norman, A. P., Ed., Blackwell Scientific, Oxford, 1971, 86.
3. **Laurence, K. M.**, Pathology of hydrocephalus, *Ann. R. Coll. Surg.*, 24, 388, 1959.
4. **Laurence, K. M.**, The effect of early surgery for spina bifida on survival and quality of life, *Lancet*, 1, 301, 1974.
5. **McLaughlin, J. F. and Shurtleff, D. B.**, Management of the newborn with myelodysplasia, *Clin. Pediatr.*, 18, 463, 1979.
6. **Althouse, R. and Wald, N.**, Survival and handicap of infants with spina bifida, *Arch. Dis. Child.*, 55, 845, 1980.
7. **Laurence, K. M., Bligh, A. S., Evans, K. T., and Shurtleff, D. B.**, Vertebral abnormalities in parents and sibs of cases of spina bifida and anencephaly, *Proc. 13th Int. Congr. Paediatr.*, 5, 415, 1971.
8. **Carter, C. O., David, P. A., and Laurence, K. M.**, A family study of major central nervous system malformation in South Wales, *J. Med. Genet.*, 5, 81, 1968.
9. **Cohen, T., Stern, E., and Roseman, A.**, Sib risk of neural tube defect: is prenatal diagnosis indicated in pregnancies following the birth of a hydrocephalic child, *J. Med. Genet.*, 16, 14, 1979.
10. **Carter, C. O. and Evans, K.**, Spina bifida and anencephaly in greater London, *J. Med. Genet.*, 10, 209, 1973.
11. **Williamson, E. M.**, Incidence and family aggregation of major congenital malformations of the central nervous system: a survey of 100 families in Southampton, *J. Med. Genet.*, 2, 161, 1955.

12. **Laurence, K. M.**, Genetics and prevention of neural tube defects in *Principles and Practice of Medical Genetics*, Vol.1, Emery, A. E. H. and Remoin, D. L., Eds., Churchill Livingston, Edinburgh, 1983,231.

13. **Nance, W. E.**, Anencephaly and spina bifida: a possible example of cytoplasmic inheritance, *Nature (London)*, 224, 373, 1969.

14. **Laurence, K. M. and Beresford, A.**, Fifty-one adults with spina bifida: continence friends, marriage and children, Dev. Med. Child Neurol., 17(Suppl.), 35, 123, 1975.

15. **Carter, C. O.**, Clues to the aetiology of neural tube malformations, Dev. Med. Child Neurol., 16(Suppl.), 32, 3, 1974.

16. **Laurence, K. M.**, Neural tube defects: a two-pronged approach to primary prevention, *Pediatrics*, 70, 648, 1982.

17. **Hale, F.**, Pigs born without eyeballs, *J. Hered.*, 24, 105, 1933.

18. **Giroud, A. and Martinet, M.**, Teratogenese par hypervitaminose A chez le rat, la souris, le cobaye et le lapin, *Arch. Fr. Pediatr.*, 16, 1959.

19. **Seller, M. J., Embury, S., Polani, P. E., and Adinolphi, M.**, Neural tube defects in curly-tail mice. II. Effect of maternal administration of vitamin A, *Proc. R. Soc. Lond.*, 206, 95, 1979.

20. **Marin-Padilla, M. and Ferm, V. H.**, Somite necrosis and developmental malformations induced by vitamin A in the golden hamster, *J. Embryol. Exp. Morphol.*, 13, 1965.

21. **Palludan, B.**, Swine in teratological studies, in *Swine in Biomedical Research*, Bustad, L. K. and McClella, R. O., Eds., Frayn Printing, Seattle, 1966.

22. **Tuchmann-Duplessis, H., Lefebvres-Boisselot, J., and Mercier-Parot, L.**, L'action tératogène de l'acide x-methyl-folique sur disease éspeces animales, *Arch. Fr. Pediatr.*, 15, 509, 1959.

23. **Nelson, M. M., Wright, H. V., Asling, C. W., and Evans, H. M.**, Multiple congenital abnormalities resulting from transitory deficiency of pteroylglutamic acid during gestation in the rat, *J. Nutr.*, 56, 349, 1955.

24. **Cheng, D. W., Bairnson, T. A., Rao, A. N., and Subbammal, S.**, Effect of variations of rations on the incidence of teratogeny in vitamin E deficient rats, *J. Nutr.*, 71, 54, 1960.

25. **Hook, E. B., Healy, K. M., Niles, A. M., and Skalko, R. G.**, Vitamin E: a teratogen or antiteratogen? *Lancet*, 1, 809, 1974.

26. **Davis, S. D., Nelson, T., and Shepard, T. H.**, Teratogenicity of vitamin B(6) deficiency: omphalocele skeletal and neural defects, and splenic hypoplasia, *Science*, 169, 1329, 1970.

27. **Woodard, J. C. and Newberne, P. M.**, Relation of vitamin B12 and one-carbon metabolism to hydrocephalus in the rat, *J. Nutr.*, 88, 37, 1966.

28. **Warkany, J.**, Effect of maternal rachitogenic diet on skeletal development of young rat, *Am. J. Dis. Child.*, 66, 511, 1943.

29. **Friedman, W. F. and Mills, L. F.**, The relationship between vitamin D and the craniofacial and dental anomalies of the supravalvular aortic stenosis syndrome, *Pediatrics*, 43, 12, 1969.

30. **Momose, Y., Akiyoshi, S., Mori, K., Nishimura, N., Fujishima, H., Imaizumi, S., and Agata, I.**, On teratogenicity of vitamin E, *Rep. Dep. Anat. Mie Prefectural Univ. School Med.*, 20, 27, 1972.

31. **Curley, F. J., Ingalls, T. H., and Zappasodi, P.**, 6-aminonicotinamide-induced skeletal malformations in mice, *Arch. Environ. Health*, 16, 309, 1968.

32. **Averback, P.**, Anencephaly associated with megavitamin therapy, *Can. Med. Assoc. J.*, 114, 995, 1976.

33. **Bernhardt, I. B. and Dorsey, D. J.**, Hypervitaminosis A and congenital renal anomalies in the human infant, *Obstet. Gynecol.*, 43, 750, 1979.

34. **Ellwood, M. J. and Ellwood, J. H.**, *Epidemiology of Anencephalus and Spina Bifida*, Oxford University Press, Oxford, 1980.

35. **Thiersch, J. B.**, Therapeutic abortions with folic acid antagonist amniopteroyglutamic acid (4-amnio-PGA) administered by oral route, *Am. J. Obstet. Gynecol.*, 63, 1298, 1952.

36. **Naggan, L. and MacMahon, B.**, Ethnic differences in the prevalence of anencephaly and spina bifida in Boston, *N. Engl. J. Med.*, 277, 1119, 1967.

37. **Naggan, L.**, Anencephaly and spina bifida in Israel, *Pediatrics*, 47, 577, 1971.

38. **Laurence, K. M., Carter, C. O., and David, P. A.**, The major central nervous system malformations in South Wales. XI. Pregnancy factors, seasonal variations and social class effects, *Br. J. Prev. Soc. Med.*, 22, 212, 1968.

39. **Leck, I.**, Further tests of the hypothesis that influenza in pregnancy causes malformations, *HSMHA Health Rep.*, 86, 166, 1971.

40. **Leck, I.**, The etiology of malformations; insights from epidemiology, *Teratology*, 5, 303, 1972.

41. **McCance, R. A., Widdowson, E. M., and Verdon-Roe, C. M.**, Study of English diets by individual methods' pregnant women at different economic levels, *J. Hyg.*, 38, 586, 1938.

42. **MacMahon, B. and Yen, S.**, Unrecognized epidemic of anencephaly and spina bifida, *Lancet*, 1, 31, 1971.

43. **Stein, Z. and Susser, M.**, Maternal starvation and birth defects, in *Birth Defects: Risks and Consequences*, Kelly, S., Hook, E. B., Janerich, D. T., and Parker, I. H., Eds., Academic Press, New York, 1976, 205.

44. **Wynn, M. and Wynn, A.,** *Prevention of Handicap and the Health of Women,* Routledge & Kegan Paul, London, 1979.

45. **Bradshaw, J., Weale, J., and Weatherall, J.,** Congenital malformations of the central nervous system, *Popul. Trends,* 19, 13, 1980.

46. South Wales Anencephaly and Spina Bifida Study Group, Preliminary report, 1981.

47. National Food Survey Committee, Annual Reports (1955-79), Household Food Consumption and Expenditure, Ministry of Fisheries and Food, Her Majesty's Stationary Office, London, 1981.

48. **Nishimura, H.,** Prenatal versus postnatal malformations based on the Japanese experience on induced abortions in the human being, in Ageing Gametes: Their Biology and Pathology, Blandau, R. J., Ed., S. Karger, Basel, 1975, 349.

49. **Burke, B. S.,** The dietary history as a tool in research, *J. Am. Diet. Assoc.,* 23, 1041, 1947.

50. **Smithells, R. W., Ankers C., Carver, M. E., Lennon, D., Schorah, C. J., and Sheppard, S.,** Maternal nutrition in early pregnancy, *Br. J. Nutr.,* 38, 497, 1977.

51. **Smithells, R. W., Sheppard, S., and Shorah, C. J.,** Vitamin deficiencies and neural tube defects, *Arch. Dis. Child.,* 51, 944, 1976.

52. **Hibbard, E. D. and Smithells, R. W.,** Folic acid metabolism and human embryopathy, *Lancet,* 1, 1254, 1965.

53. **Laurence, K. M., James, N., Miller, M., and Campbell, H.,** Increased risk of recurrence of neural tube defects to mothers on a poor diet and possible benefits of dietary counseling, *Br. Med. J.,* 281, 1542, 1980.

54. **Laurence, K. M., James, N., Miller, M., Tennant, G. B., and Campbell, H.,** Double-blind randomised controlled trial of folate treatment before conception to prevent recurrence of neural tube defects, *Br. Med. J.,* 282, 1509, 1981.

55. **Tennant, G. B. and Withey, J. L.,** An assessment of work simplified procedures for the microbiological assay of serum vitamin B12 and serum folate, *Med. Lab. Technol.,* 29, 171, 1972.

56. **Hoffbrand, A. V., Newcombe, B. F. A., and Mollin, D. L.,** Method of assay of red cell folate activity and the value of the assay as a test for folate deficiency, *J. Clin. Pathol.,* 19, 17, 1966.

57. **James, N., Laurence, K. M., and Miller, M.,** Diet as a factor in the aetiology of neural tube malformation, *Z. Kinderchir. Grenzgeb.,* 31, 302, 1980.

58. **James, N. and Laurence, K. M.,** Nutrition and the prevention of neural tube defects, in *Topics in Paediatric Nutrition,* Dodge, J. A., Ed., Pitman, Tunbridge Wells, in press.

59. **Laurence, K. M. and Campbell, H.,** Folate treatment to prevent recurrence of neural tube defect, *Br. Med. J.,* 282, 2131, 1981.

60. **James, N.,** Personal communication, 1982.

61. **Smithells, R. W., Sheppard, S., Schorah, C. J., Seller, M. J., Nevin, N. C., Harris, R., Read, A. P., and Fielding, D. W.,** Possible prevention of neural tube defects by periconceptional vitamin supplementation, *Lancet,* 1, 339, 1980.

62. **Smithells, R. W., Sheppard, S., Schorah, C. J., Seller, M. J., Nevin, N. C., Harris, R., Read, A. P., and Fielding, D. W.,** Apparent prevention of neural tube defects by periconceptional vitamin supplementation, *Arch. Dis. Child.,* 56, 911, 1981.

63. **Smithells, R. W.,** Neural tube defects: prevention by vitamin supplements, *Pediatrics,* 69, 498, 1982.

64. **Laurence, K. M.,** Spina bifida research in Wales, *J. R. Coll. Phys.,* 10, 333, 1976.

65. **Laurence, K. M.,** Towards prevention of neural tube defects, *J. R. Soc. Med.,* 75, 728, 1982.

66. **Laurence, K. M.,** Neural tube defects: a two pronged approach to primary prevention, *Pediatrics,* 40, 248, 1982.

67. **Schorah, C. J., Smithells, R. W., and Scott, J.,** Vitamin B12 and anencephaly, *Lancet,* 1, 880, 1980.

68. **Beck, F.,** Comparative placental morphology and function, in *Developmental Toxicology,* Kimmel, C. A. and Buelke, J., Eds., Raven Press, New York, 35, 1981.

69. **Tanimura, T. and Shepard, T. H.,** Glucose metabolism by rat embryo *in vitro, Proc. Soc. Exp. Biol. Med.,* 35, 51, 1970.

70. **Shepard, T. H. and Tanimura, T.,** Energy metabolism in early mammalian embryos, *Dev. Biol.,* 4(Suppl.) 42, 1970.

71. **Watt, D. J.,** Mechanism of the Development of Spina Bifida in the Chick, Ph.D. Thesis, Aberdeen University, 1977.

72. **Wilson, J. G.,** *Environment and Birth Defects,* Academic Press, New York, 1973.

73. **New, D. A. T.,** Whole embryo culture and the study of mammalian embryos during organogenesis, *Biol. Rev.,* 53, 81, 1978.

74. **Elmes, M. E.,** Zinc absorption and metabolism in *Paediatric Nutrition,* Dodge, J. A., Ed., Pitman, Tunbridge Wells, in press.

75. **Bemmer, I. and Mills, C. F.,** Absorption, transport and tissue storage of essential trace elements, *Phil. Trans. R. Soc. Lond. B,* 294, 75, 1981.

76. **Tamura, T., Shane, B., Boer, M. T., King, J. C., Margen, S., and Stokstad, E. L. R.,** Absorption of mono- and poly-glutamyl folates in zinc depleted man, *Am. J. Clin. Nutr.,* 31, 1984, 1978.

77. **Shojania, A. M. and Hornaby, G. J.,** Oral contraceptives and folate absorption, *J. Clin. Lab. Med.,* 82, 869, 1973.
78. Food and Nutrition Board, Recommended Dietary Allowances, Publ. 1694, 7th ed., National Academy of Sciences—National Research Council, Washington, D.C., 1968, 35.
79. **Edwards, J. H.,** Vitamin supplementation and neural tube defects, *Lancet,* 1, 648, 1982.
80. **Seller, M. J.,** Nutritional supplementation and prevention of neural tube defect, in *Clinical Genetics: Problems in Diagnosis and Counseling,* Willey, A. M., Carter, T. P., Kelly, S., and Porter, I. H., Eds., Academic Press, New York, 1982, 1.
81. **Stone, D.,** Possible prevention of neural tube defects by periconceptional vitamin supplementation, *Lancet,* 1, 647, 1980.
82. **Laurence, K. M.,** Prevention and prenatal diagnosis of neural tube defects, in *Advances in the Study of Birth Defects, Vol. 7, Central Nervous System and Craniofacial Malformations,* Persaud, T. V. N., Ed., MTP Press, Lancaster, 1982, 133.
83. **Laurence, K. M.,** Antenatal detection of neural tube defects, in *Fetal and Neonatal Pathology,* Barson, A. J, Ed., Prager, Eastbourne, 1982, 7.
84. **Laurence, K. M.,** The prenatal diagnosis of anencephaly and spina bifida, in *Birth Defects: Risks and Consequences,* Kelly, S., Janerish, D. T., and Hook, I. H., Eds., Academic Press, New York, 1976, 294.
85. **Laurence, K. M., Campbell, H., and James, N.,** The role of maternal diet and preconceptional folic acid supplementation in the prevention of neural tube defects, in **Prevention of Spina Bifida and Other Neural Tube Defects,** Dobbing, J., Ed., Academic Press, London, 1983, 85.

INDEX

T - #0659 - 101024 - C0 - 253/174/12 - PB - 9781138561663 - Gloss Lamination